Grasslands - Conservation and Development

Edited by Muhammad Aamir Iqbal

Published in London, United Kingdom

Grasslands - Conservation and Development
http://dx.doi.org/10.5772/intechopen.107797
Edited by Muhammad Aamir Iqbal

Contributors
Ally Kiyenze Nkwabi, Asahi Sakuma, Baraka Naftal Mbwambo, Bukombe John Kija, Chanjia Li, Emmanuel Pagiti Reuben, Hiroto Shimazaki, Luis Fernando Galeano-Vasco, Manuela Ortega Monsalve, Mario Fernando Ceron-Munoz, Marisol Medina-Sierra, Muhammad Aamir Iqbal, Nidaa Harun, Noordeen Nusrathali, Pius Yoram Kavana, Rahmathulla Mohamed Nikzaad, Raina Ijaz, Saima Iqbal, Satoshi Irei, Satoshi Kameyama, Seiichiro Yonemura, Selemani Rehani Moshi, Serap Kizil Ayedmir, Shabani Matwili, Simula Peres Maijo, Stephen Justice Nindi, Tatiana Rodríguez Monroy, Victor Alexander Kakengi, Xiuqin Lin

First published in London, United Kingdom, 2024 by IntechOpen
IntechOpen is the global imprint of INTECHOPEN LIMITED, registered in England and Wales, registration number: 11086078, 167-169 Great Portland Street, London, W1W 5PF, United Kingdom

British Library Cataloguing-in-Publication Data
A catalogue record for this book is available from the British Library

Additional hard and PDF copies can be obtained from orders@intechopen.com

Grasslands - Conservation and Development
Edited by Muhammad Aamir Iqbal
p. cm.

This title is part of the Agricultural Sciences Book Series, Volume 8
Topic: Agronomy and Horticulture
Series Editor: W. James Grichar
Topic Editor: Ibrahim Kahramanoglu
Associate Topic Editors: Murat Helvaci and Olga Panfilova

Print ISBN 978-0-85014-123-8
Online ISBN 978-0-85014-124-5
eBook (PDF) ISBN 978-0-85014-125-2
ISSN 3029-052X

For EU product safety concerns:
IN TECH d.o.o., Prolaz Marije Krucifikse Kozulić 3, 51000 Rijeka, Croatia,
info@intechopen.com or visit our website at intechopen.com.

IntechOpen Book Series

Agricultural Sciences

Volume 8

Aims and Scope of the Series

The importance of agriculture cannot be overstated. It helps sustain life, as it gives us the food we need to survive and provides opportunities for economic well-being. Agriculture helps people prosper around the world and combines the creativity, imagination, and skill involved in planting crops and raising animals with modern production methods and new technologies. This series includes two main topics: Agronomy and Horticulture, and Animal Farming. This series will help readers better understand the intricacies of production agriculture and provide the new knowledge that is required to be successful. The success of a farmer in modern agriculture requires knowledge of events happening locally as well as globally that impact input decisions and ultimately determine net profit.

Meet the Series Editor

W. James Grichar has been employed with Texas A&M AgriLife Research for over 45 years with an emphasis on research in agronomy, plant pathology, and weed science. He obtained his BS from Texas A&M in 1972 and his Masters of Plant Protection in 1975. He has published 195 journal articles, over 330 research reports and briefs, 11 book chapters, and over 300 abstracts of profession meetings. He also directs research in many crops including corn, grain sorghum, peanuts, and sesame. He has held various positions in different professional societies including the American Peanut Research and Education Society, Southern Weed Science Society, and Texas Plant Protection Conference in addition to being Associate Editor for Peanut Science and Weed Technology. Significant accomplishments have included spearheading efforts to determine the optimum planting time for soybean production along the upper Texas Gulf Coast. These efforts have shown growers that soybean yields can be improved by 10 to 20% by following a late March to early April plant date. He also has been instrumental in developing a herbicide program for peanut production in the south Texas growing region. Through the development and use of herbicides that are effective against major weed problems in the south Texas region, peanut yields have increased by 25 to 30%.

Meet the Volume Editor

Dr. Muhammad Aamir Iqbal, Ph.D., is an academician and researcher in agronomy with research interests in sustainable crop production, genetic improvement of field crops, organic farming, seed science, grasslands and watershed management, and bioengineering-based solutions for converting crop residues into useful products. Dr. Iqbal is currently engaged as a Fulbright postdoctoral scientist at Louisiana Tech. University, USA, having a role in advancing the scientific research on leftover crops management. Previously, he completed research training at the University of California Davis, USA, Shandong Academy of Agricultural Machinery Sciences, China, and Yunnan Sugarcane Research Institute, China. Dr. Iqbal was conferred a Doctorate in Agronomy by a prestigious seat of agricultural learning and research in South Asia, University of Agriculture Faisalabad, Pakistan. He has more than 143 scientific publications and dozens of book chapters on diversified aspects of sustainable crop production and genetic improvement of field crops to his credit. Additionally, he has edited three books on different aspects of crop production, weed management, and grassland conservation. Dr. Iqbal is an editorial board member and reviewer for many high-impact journals. Being a member of the National Curriculum Revision Committee (NCRC) of Pakistan's Higher Education Commission (HEC), he has been involved in the critical task of revising and updating the curriculum of undergraduate and postgraduate degree programs offered by Pakistani Higher Education Institutes in the subjects of agronomy and crop physiology. Dr. Iqbal is a faculty member in the Department of Agronomy, Faculty of Agriculture, University of Poonch Rawalakot, Azad Jammu and Kashmir, Pakistan, having responsibilities of teaching, research, research mentorship, administration, and outreach activities. Moreover, he has completed research projects as principal investigator and supervised the research of several postgraduate scholars.

Contents

Preface

Globally, the attainment of the United Nations Sustainable Development Goals (SDGs) established in 2015, especially poverty alleviation, zero hunger, clean water and sanitation, decent work and economic growth, climate action, and life on land, are both directly and indirectly associated with grasslands. The underlying reason is that grasslands, also referred to as prairies, savannas, steppes, and meadows, play a strategic role in sustaining diverse ecosystems and provide a wide array of crucial benefits to the environment, wildlife, livestock, and human communities. For instance, grasslands play an important role in carbon sequestration wherein the grasses store bulk quantities of carbon in the soil through their deep root systems. Grasslands are also significant in sustaining biodiversity (plants, birds, and animal species), controlling floods, providing water through effective regulation of rainwater, and preventing erosion and improving infiltration and hotspots of recreational activity such as hiking, birdwatching, and so on. Moreover, grasslands are precious sources of feed for livestock, herbs for traditional medicines, and rearing spots for honeybees. They also have cultural value in terms of spiritual and religious sites. However, grasslands have remained quite vulnerable to invasive weed species, anthropogenic disturbances (e.g., land conversion for carrying out modern input-intensive farming, uncontrolled grazing by livestock, etc.), and numerous types of climate change. This book, *Grasslands – Conservation and Development*, has been tailored to address these issues. It suggests practices for conserving grasslands that ensure the protection and sustainable management of grassland ecosystems by maintaining the biodiversity and ecological integrity for persistent provision of ecosystem services, as well as grassland development initiatives intended for improving the productivity and sustainability of grassland areas for agricultural purposes.

The book is organized into two sections. Section 1, "Novel Grasslands Conservation Approaches", synthesizes fundamental concepts, recent advancements, and knowledge on sustainable management, biologically viable protection, economic utilization, effective restoration, and viable preservation of grasslands. Section 2, "Grasslands Development Initiatives", includes chapters that discuss recently developed initiatives that ensure the conversion of grasslands into more profitable, self-sustaining, and ecologically balanced natural or improved entities.

Chapter 1 discusses the fundamentals and advancements of grassland conservation and development using artificial intelligence (AI) tools. It suggests that AI-based tools might be developed and optimized as viable, reliable, and cost-effective methods to monitor and assess grassland production potential under changing pedo-climatic scenarios. Moreover, it elaborates that digitalized monitoring using automated tools might assist in recognizing native grass species as well as invasive species and analyzing soil fertility status for formulating effective management and conservation options. For instance, unmanned aircraft systems (UAS) integrating numerous types of cameras and sensors hold the potential to precisely collect scattered spatial and temporal information with remarkably high resolution in the visible and infrared spectrum. Likewise, this chapter forwards the idea that AI, deep learning (DL), and machine learning (ML) could aid in the construction of 2D and 3D models for establishing grass species composition, total vegetation cover, barren patches, and nutritional quality of grass species. Furthermore, ML and AI algorithms may be employed for localizing grass species from data supplied by images and sensors. Using these approaches, grass species can be localized down to the centimeter scale and maps can be constructed to illustrate grass cover distribution on large swathes of grasslands.

Chapter 2 examines grasslands in terms of carbon sink, ecological barriers and watershed for low riparian regions, feedstock, mineral extraction sites for drilling and mining, and numerous associated benefits like wool, herbs for traditional medicines, tourism and leisure, and more in China and Pakistan. Additionally, it highlights that grassland ecosystems have been persistently degraded by anthropogenic disturbances (land use changes, tourism, intensive grazing, fire, vegetation clearance, invasive weeds such as knapweed, parthenium, Johnsongrass, etc.) and climate change drivers (heat, drought, chilling, salinity, and shifting of rainfall patterns) in both countries. Moreover, it suggests that for conserving grasslands, nitrogen (N) and carbon (C) in these countries' soils are integral for maintaining the primary productivity of grass species; hence estimating the extent of numerous interventions on N and C cycling along with grass–microbe interactions become imperative from socioeconomic and environmental perspectives. The chapter also compiles recent knowledge on the productivity status and persistent degradation of grasslands in China and Pakistan. Finally, the chapter analyzes the prevalence of invasive weeds in grasslands, grass–microbe interactions and their effect on the growth of plant species, microclimate, and availability of nutrients as well as synthesizes recent advances in C and N dynamics in grasslands ecosystems.

Chapter 3 proposes that grasslands in lowlands and mountains either in natural form or developed landscape can provide an added value in terms of ecotourism opportunities, owing to having huge esthetic and recreational potential compared to uniform agricultural areas. It has been objectively highlighted that grasslands characterized by high species and habitat diversity-based ecotourism are a form of nature-based tourism whereby people visit natural or developed areas for recreation, sightseeing, permitted and controlled hunting, on-site purchase of organic products, and more, and are usually managed by adopting sustainable practices. Moreover, it highlights that ecotourism can generate multifaceted economic advantages for local communities such as the direct sale of products to tourists. This chapter highlights a variety of negative impacts such as the high number of tourists, which leads to overuse of resources. Finally, the authors describe pronounced challenges to the development of grasslands for ecotourism, such as lack of community cooperation, careless herders, need of hefty investment, and absence of trained human capital along with climate change and loss of biodiversity.

Chapter 4 reports that prescribed grassland burning in Japan (commonly referred to as Noyaki in Japanese) for the purpose of grassland conservation has been conducted for over a thousand years by local residents. It highlights that biomass burning has emerged as one of the most significant sources of airborne substances including mercury. Additionally, during Noyaki, the chapter authors sampled and measured gaseous mercury from the Noyaki plumes to better understand mercury emission and cycle in the local environment. The chapter reports on the remarkable elevation of atmospheric mercury concentrations, demonstrating the emission of gaseous mercury from the Noyaki. Lastly, the chapter discusses possible origins, novel information on the results inferred, and future research directions.

Chapter 5 highlights that many conventional farming approaches in developing nations segregate livestock and crop production, ignoring the synergistic advantages and sustainable land management possibilities that may be gained by combining the two. It suggests that in order to increase agricultural output and foster grassland conservation, the idea of merging livestock and crop systems must be considered. By highlighting the advantages and challenges of the approach, this chapter draws attention to the potential benefits of integration, including enhanced soil fertility, efficient resource use, increased productivity, and better protection of grassland ecosystems. Moreover, the chapter explores the importance and viability of a variety of integrated agricultural methods, including agro-pastoral, mixed, and silvopastoral systems, in a variety of geographical settings. Finally, this chapter culminates with solid suggestions to educate policymakers, academics, and practitioners about the need to integrate livestock and crop production to achieve long-term agricultural sustainability in low-income nations.

Chapter 6 focuses on determining the quality of grasses using spectroscopy techniques, specifically, spectrophotometry. It points out that chemical methods used to determine the nutritional quality of grasses produce chemical residues, are time-consuming, and are costly to use when analyzing large crop extensions. Thus, the chapter authors suggest that spectroscopy, which is a non-destructive technique, can establish the nutritional quality of grass easily and accurately. This chapter describes the techniques focused on the use of spectroscopy and machine learning models to predict and determine the quality of grasses. Moreover, it compiles different methods and results focusing on the quality of the grasses. In general, it shows that the most used spectroscopic method is the near-infrared analysis. In addition, the statistical model most used by the authors to predict the quality of grasses is the partial least squares regression. Moreover, this chapter interestingly portrays spectroscopy as an effective tool that opens the way to new types of technologies for determining the quality of grasses.

Chapter 7 provides an overview of the fundamental aspects of cation exchange capacity (CEC) and its significance as a soil fertility indicator, particularly in evaluating pasture quality. It highlights the limitations of conventional methods and emphasizes the need to explore alternative approaches for rapid pasture quality assessment. In addition, this contribution focuses on the growing importance of hyperspectral information (HSI) as a valuable tool for efficient and accurate CEC estimation. It also discusses the advantages and challenges associated with using HSI data for CEC estimation and its potential to enhance decision-making in pasture and forage production. This chapter highlights that HSI technology allows for the capture and analysis of a wide spectrum of light reflectance information at various wavelengths, providing valuable insights into the physical and chemical characteristics of the soil.

Chapter 8 delves into the intricate relationship between agro-pastoralism and grassland ecosystems in Tanzania's Western Serengeti and Ugalla ecosystems. This chapter reveals that despite the acknowledged contribution of agro-pastoralism to rural well-being and economic development, its impact on the delicate balance of grassland ecosystems remains unclear in these crucial Tanzanian landscapes. Moreover, it illuminates agro-pastoralism's environmental, social, and economic dimensions in these regions. Guided by research questions exploring current conditions, potential solutions, and the path toward sustainable grassland resource utilization, the study employs a systematic approach to develop grasslands for promoting agro-pastoralism. Furthermore, it highlights challenges pertaining to the progressive expansion of agro-pastoral activities, leading to trade-offs between ecosystem services and productivity. Finally, this chapter concludes with the identification of agro-pastoral clusters across the area, revealing variations in economic activities and their impact on grassland utilization.

Key Features of the book:

- Furnishes fundamental and state-of-the-art knowledge on grassland conservation using artificial intelligence tools.

- Presents integrated strategies to impart balance in carbon-nitrogen cycles in grasslands to ensure the food security of future generations.

- Provides vital information on grassland development for ecotourism.

- Describes and discusses the role of soil microbes in grasslands.

- Highlights the importance of soil fertility indicators for conserving grassland soil.

- Illustrates potential strategies to assess the nutritional quality of grass species.

- Depicts recent knowledge on integrating crop-livestock for poverty alleviation and attaining zero hunger.

I would like to acknowledge the strategically vital and pertinent intellectual support furnished by my mentor Dr. Asif Iqbal (Associate Professor, Department of Agronomy, University of Agriculture Faisalabad, Pakistan) and Dr. Chunjia Li (Sugarcane Research Institute, Yunnan Academy of Agricultural Sciences, China).

Muhammad Aamir Iqbal
Faculty of Agriculture,
Department of Agronomy,
University of Poonch Rawalakot,
Azad Jammu & Kashmir, Pakistan

Section 1

Novel Grasslands Conservation Approaches

Chapter 1

Introductory Chapter: Present and Future of Artificial Intelligence in Grasslands Conservation

Muhammad Aamir Iqbal

1. Introduction

The grasslands entail any ground cover encompassing grasses as a predominant vegetation, whereas shrubs and tree cover generally tend to prevail on less than 10% of that area [1, 2]. Recently, grasslands conservation and development have attained the attention of researchers and policy makers owing to their strategic pertinence in terms of carbon sink, watershed areas, feed source for livestock etc. Fundamentally, the concept of grasslands conservation involves those initiatives which ensure protection of grassland ecosystems and sustainable provision of ecosystem services through scientific maintenance of their ecological integrity and preventing the loss of biodiversity. Besides this, there is another concept of grasslands development which encompass different anthropogenic interferences (e.g., construction of roads, dams, wild-life sanctuaries, amusement parks etc.) intended for diversifying and multiplying the ecosystems services offered by grassland areas primarily through enhancement of their agricultural usage [3]. Traditional grassland ecosystem monitoring has mainly relied on field surveys, however these are being increasingly enriched and developed towards a large range, high spatio-temporal resolution and high-precision directions. The conservationists and researchers call for innovative solutions to effectively manage a variety of anthropogenic interferences and handle environmental problems that have recently seriously threatened the ecosystem functioning of grasslands.

Grasslands conservation requisites the exploration of their productive potential in order to plan grazing and reseeding activities [1, 2]. Currently, approaches employed for grasslands monitoring (e.g., visual assessments) are labour demanding, costly and, therefore have remained inadequate in terms of practical use. Additionally, nutritional value of native grass species has remained neglected and un-monitored since their quality assessment requires intricate laboratory analyses, involving huge expenditures and technical expertise [4]. Alternatively, artificial intelligence (AI) based tools might be developed and optimized as viable, reliable and cost-effective methods to monitor and assess grassland production potential under changing pedo-climatic scenarios [5]. Moreover, digitalized monitoring using automated tools might assist in recognizing native grass species, invasive species and analysing soil fertility status for formulating effective management and conservation options. For instance, unmanned aircraft systems (UAS) integrating numerous types of camera and sensors hold potential to precisely collect scattered spatial and temporal information having remarkably high-resolution in the visible and infrared spectrum [6]. Likewise, AI, deep learning (DL) and machine learning (ML) could aid to construct 2D and 3D models [7] for establishing grass species composition, total vegetation cover, barren patches, and nutritional quality of grass species. Furthermore, ML and AI algorithms may be employed for localizing the grass species from the images and sensors supplied data. Using these approaches, grass species can be localized down to centimetre scale and allows maps construction to illustrate grass cover distribution on large swathes of grasslands.

This chapter briefly introduces fundamentals and application types of artificial intelligence with special emphasis on deep learning, drone's types and their utilization potential for grasslands monitoring and conservation. Moreover, different challenges that could emerge while using artificial intelligence tools for grasslands conservation and development have also been objectively highlighted.

2. Artificial intelligence (AI)

The AI entails an amalgamation of technologies having human-like cognitive capacities which enable them to precisely learn, accurately perform, and effectively make decisions. It is fundamentally a software which might initiate logical reasoning, learn from patterns, and resultantly can solve intricate problems in a highly rationale manner [8, 9]. Interestingly, AI is typically defined in a specific computer science method's context that it uses such as machine learning (ML), reinforcement learning (RL), and deep learning (DL). The AI software can also be physically implemented in the form of humanoid robots, autonomous cars and robotic hands. The presence of AI is felt wherever any type of machine is performing a wide range of tasks by utilizing reasonably high level of independent intelligence that is typically practiced by humans [8, 10]. In comparison to natural intelligence of humans and other biological organisms, the machine intelligence is different in the sense that it is artificially created and digital in nature and functioning. In recent generation AI has emerged as one of the most exciting research area owing to its unprecedented potential to solve the intricate problems particularly to reduce labour use and ensure efficient resource utilization [8, 9, 11]. In agriculture, a wide range of AI approaches have been suggested including robotics, DL, ML, neural network (NN), support vector machine (SVM), random forest etc. **Figure 1** illustrates AI applications especially cognitive related applications (data fusion and mining, artificial neural networks etc.), robotics applications (unmanned aerial vehicles or drones, swarm robots etc.) and natural interface concerning applications (virtual and augmented reality, computer vision etc.).

Broadly, intelligent systems are classified into four categories including (i) systems having humans like cognitive potential, (ii) systems capable to act like humans, (iii) systems that rationally think to solve any problem, and (i*ν*) systems which act rationally to solve intricate challenges [9, 12]. This classification of intelligent systems is based on their potential to think and act while their success is measured in comparison to human rationality and performance. Thus, AI systems hold potential to assist in grasslands conservation as these can be effectively used for manipulation and long-term storage of data regarding grass cover and composition along with soil condition and pest record. Data manipulation entails potential to deduce and infer novel knowledge from existing information and data set. Additionally, these AI systems can precisely record data through vigorous acquisition of evaluation parameters at inaccessible altitudes of grasslands along with data analyses and representation through interpretation of the acquired knowledge. Among AI techniques, artificial neural

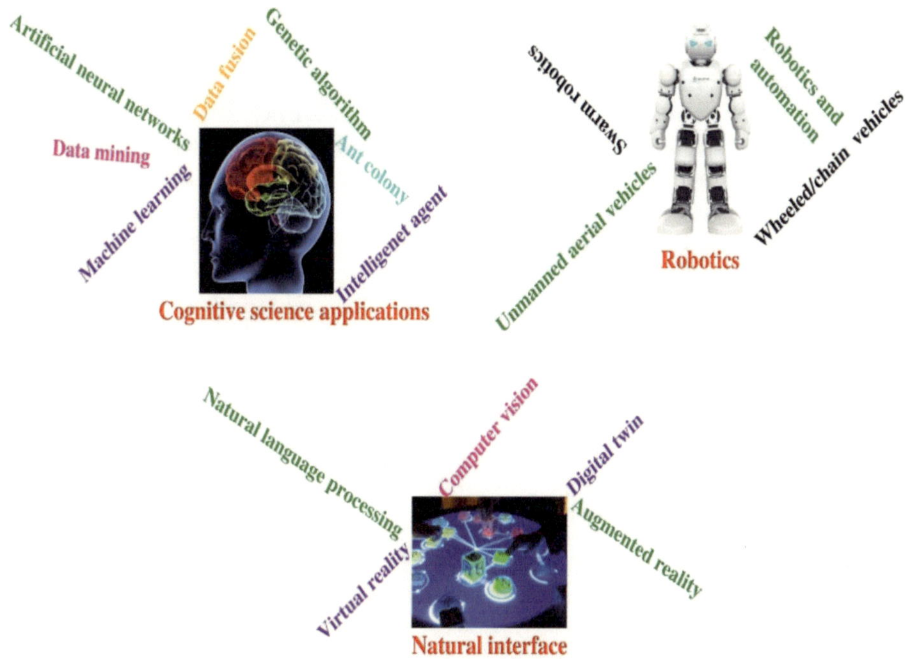

Figure 1.
Different types of artificial intelligence applications especially cognitive applications (data fusion and mining, artificial neural networks etc.), robotics applications (unmanned aerial vehicles or drones, swarm robots etc.) and natural interface applications (virtual and augmented reality, computer vision etc.).

networks (ANN) has emerged as one the most strategic technique that is developed by interconnecting nodes having potential to perform intricate functions as human brain [10, 13, 14]. This technique has been successfully employed in agriculture for estimating and forecasting methane emission from fields and soil moisture contents. Besides, scaled conjugate gradient and Quasi-Newton based neural network algorithms have been employed to estimate soil traits and environmental variables impact and predict dynamics of soil moisture content on hourly basis [12, 14]. However, focus of this study is robotics, deep learning and drones for their potential utilization in grasslands conservation and development.

3. Robotics

Robotics a branch of technology aims to design and construct robots for different applications and operations. Robots are basically machine resembling humans having potential to replicate actions typically performed by humans [8, 9]. There have been industrial robots and mobile robots during last many decades and now service robots have been focussed in order to develop their closer interaction with humans to fulfil their emerging needs. Few pertinent examples of service robots include medical robots, underwater robots, field robots, rehabilitation robots, construction robots and humanoid robots [9]. Among AI approaches, robots are currently most widely used in agriculture after drones to assist in many operations including visual detection, weeding, spray application, and harvesting having flexibility and adjustment options to match the requirement of various tasks. Such robots are typically two-wheeled entailing a mobile base to execute a spewing mechanism to control the movements of the robot through a wireless tool. Additionally, robots might be fitted with high resolution cameras to monitor the growth conditions of native grass species and monitor the effects of abiotic stresses (drought, heat, chilling, salinity, heavy metal toxicity, soil compaction, water logging etc.). Moreover, this monitoring can effectively reveal and detect the presence and extent of pest attack in different areas of grasslands. Similarly, robots can be equipped to collect soil samples for determining the physicochemical characteristics of the grasslands soils at varying altitudes which are typically inaccessible for humans. Moreover, robots can perform the function of surveillance for reporting over-grazing and invasion of exotic species in the grasslands which in the longer run results in the deterioration of ecosystem functioning of the grasslands. However, research and practical implementation of robots based surveillance programs are still awaited for grasslands monitoring and conservation.

4. Deep learning (DL)

The DL technology is relevant to AI, ML and data science (DS) with advanced analytics and recently it has emerged as a focussed research area in computer science and intelligent computing. Generally, AI is the name of incorporating human intelligence and behaviour to different machines, set of machines or complex machine systems [13], while ML represents learning method from data set or experience, which automates the building of analytical models. Likewise, DL also entails learning from data set whereby computation is processed through multi-layer neural networks [7, 13, 15]. In DL, the term deep encompasses the concept of multiple stages or levels through which processing of data are performed in order to build a data-driven model. Therefore, DL is a technique that trains multiple computers enabling them to process information in such a way which mimics human neural processes. It is fundamentally a ML technique that instructs computers to learn by example from large data sets and architectures of neural networks. The DL technique has been successfully employed to teach computers that imitates the way humans gain and process intricate information and knowledge [13].

The original idea of DL was envisaged in 1943 [12] and it was meant for creating a model entailing a single biological neuron and its progressive development to link numerous individual neurons together to build an ANN. Recently, deep neural networks (DNN) have superseded the ANN by virtue of comparatively larger number of layers and multiple data sets. It has been enabled to separate signal of interest from noise in data set by utilizing algorithms including backpropagation in order to optimize parameters of interest in a single layer from the previous layer [13].

Recently, DL based methods have been developed to process the data set of ultra-high-resolution images which might be utilized for monitoring the grasslands features like vegetation cover and extent of exotic species invasion. Particularly, deep CNNs (convolutional neural networks) have recently revolutionized the image processing and their interpretation through remarkable improvement in specified object detection and precise classification of assigned tasks [13, 15]. Interestingly, deep CNN is basically a network which is comprised of many layers to take image's pixels as input data to predict interpretation as output. The underlying mechanism is CNNs primarily apply different learned filters (thousands in number) to the image's regions internally and finally combines those to find out targeted information [15]. Importantly, DL uses end-to-end approach which allows targeted features automatic detection without human interaction owing to its capability to learn and detect the desired features. Recently, such methods have been successfully introduced for the detection and counting of plants and animals in varying ecological context and this feature might be utilized to recognize native grass species cover in grasslands in order to formulate the organize conservation initiatives.

5. Drones

Drones are basically airborne self-propelled devices having a programmable regulator for controlling and regulating their movement without any on-board pilot. Since last decade, drones also known as unmanned aerial systems (UAS), remotely piloted aircraft systems (RPAS) or unmanned aerial vehicles (UAV) have attracted the attention of researchers in civilian and scientific spheres, because of their potential to usher a new era of precision agriculture, remote sensing and environment studies [16, 17]. The UAV can provide images of high resolution of natural phenomena at a relatively low-cost, risk-free, rapid and systematic manner [18, 19]. For these reasons, UAV hold potential to become a major trend in grasslands and wildlife related research. Drones are classified in various ways-by size, range, endurance and carrying capacity. **Figure 2** portraits the classification of different types of drones based on their size, range, take-off weight, wing type, power source, assembling requirement and nature of use or applications (monitoring or logistic). However, wing based categorization is the simplest classification especially fixed wing drones are capable

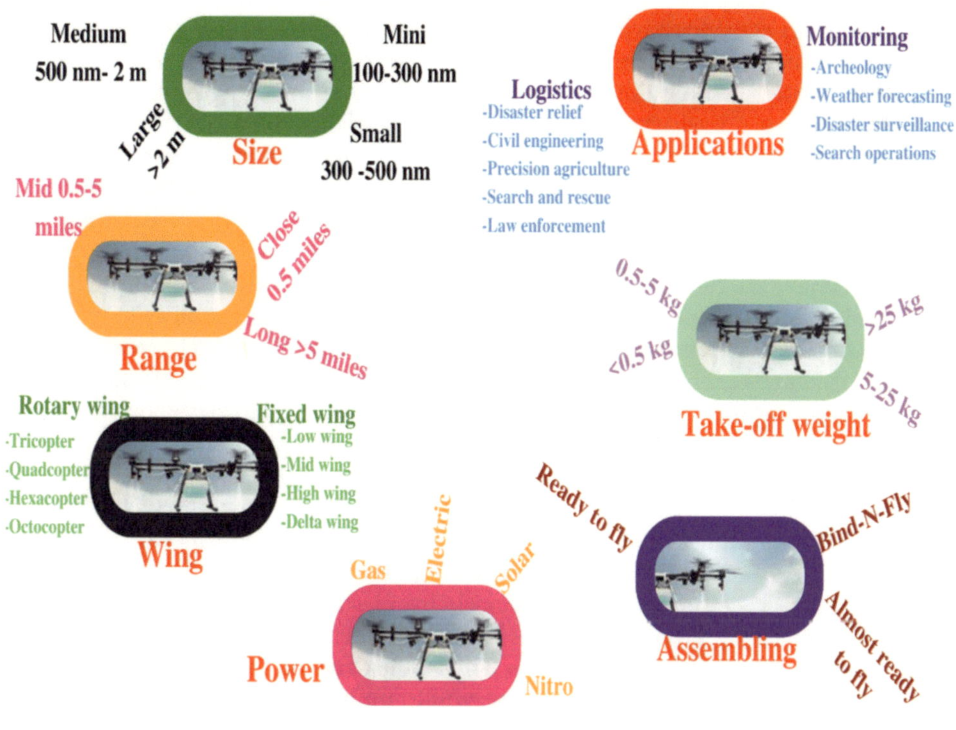

Figure 2.
Classification of different types of drones based on their size, range, take-off weight, wing type, power source, assembling requirement and nature of use or applications (monitoring or logistic).

to carry heavy loads over comparatively longer distances while working just like airplanes [16, 19–22].

Likewise, ready to fly drones have higher take-off weight potential and thus, these hold bright perspectives to perform over-seeding operation on vast swathes of grasslands. Contrarily, small sized (fixed-wing and rotary-wing) drones might find their use in generating high resolution videos and still photography of grasslands situated on varying altitudes. Such types of UAVs equipped with multispectral sensors and lightweight cameras can potentially deliver real-time professional mapping at a fraction of the cost involved in previously human-employed photogrammetric techniques. Additionally, compact thermal vision cameras might also be installed on medium size UAV along with hyperspectral sensors and laser scanners [19, 21] for vegetation studies in grasslands and forests. Moreover, drones can be equipped with sensing devices [18, 21] for recording distinct environmental attributes (temperature, relative humidity, precipitation or air pollution) of far-flung grasslands. Moreover, large size aerial platforms having capability to lift heavier payloads might represent one of the most economical solution for sampling of grasslands soils or deliver grass seeds in degraded patches.

The UAV's success in performing grassland conservation operations can be partially predicted by keeping in view their great flexibility to carry different types of cameras, sensors, lasers and other devices. The scope of operations (visual observance, species recognition, grass composition assessment, over-seeding, nutrients application, invasive species intrusion assessment etc.) might determine the optimum combination of airborne devices and payload. These UAVs can assist in accurate retrieval of grassland traits to formulate conservation plans by adjusting grazing frequency and intensity. Optical imaging systems might be carried by airborne devices as payload which might serve as an alternate of non-destructive methods for nutritional quality assessment of grass species. Similarly, vegetation spectral response by employing visible and near-infrared wavelengths can generate useful information on vegetation composition in grasslands. Moreover, by combining multivariate modelling with spectral measurements can reveal complex relationship between grass traits and canopy reflectance. The grasses canopy's spectral response in the visible and near-infrared (NIR) regions (between 450 and 900 nm wavelength) can unveil biophysical and biochemical traits such as canopy cover, chlorophyll content, primary nutrients (nitrogen, phosphorous and potassium) contents of grass species. Furthermore, UAV borne imaging systems entail capability to provide spatially continuous information, which is much more advantageous in comparison to point-based measurements for exploring spatial patterns of grasslands monitoring parameters. Drones can be equipped to estimate grasses traits based on canopy spectral measurements in the optical domain that is generally classified as empirical and physically-based approaches. Empirical approaches rely on large data set to fit prediction models, whereas physical approaches need adequate parametrization of a Radiative Transfer Model (RTM) in order to translate spectral information into grass traits. The RTM-based frameworks have greater potential for generalization however performance could be suboptimal particularly in grasslands having multiple grass species. Contrastingly, empirical models have better adaptability for complex datasets however, these have limited applications. Thus, quick and risk-free evaluation favours deployment of UAV borne optical imagery, while empirical models might be utilized as an initial option to characterize grasslands.

Traditional assessment methods (survey expeditions by technical staff or researchers) generally focus on recording data of biomass cover seasonally or occasionally which limit drawing accurate conclusions for grasslands conservation and development owing to small data sets. However, grass species in grasslands need frequent monitoring owing to changing climate scenarios, wild fires, uncontrolled intensity of grazing by non-concerned ranchers, deteriorated soil conditions, invasive plant species etc. There is dire need to develop a perennial monitoring systems for grasslands that has potential to cope with rapidly varying climatic conditions because such frequent data acquisition is vital for making informed decisions. Therefore, prediction models relying on limited spectral datasets are always prone to errors while such uncertainties can be effectively dealt with the use of aerial borne platforms equipped with high resolution cameras and state-of-the-art sensor systems.

6. Challenges and future perspectives

The AI based models hold advantages of being flexible and adaptive, however their application for monitoring grasslands can be intricate and opaque. Nonetheless, these opaque models may become computationally expensive (exponentially) with higher levels of data complexity. Despite all advantages, the ANNs are extremely data intensive and requisite time-consuming parameter tuning approaches to process information [9, 10]. These limitations of ANNs have led to the development of easier-to-train algorithms including decision trees (DTs), support vector machines (SVMs), and random forests (RFs), however their application for grasslands monitoring still awaits target-specific research. In addition, there is need to create trust on AI based forecasts along with developing ethical and moral standards of their application in grasslands conservation and development. To achieve these goals, detailed explanations of AI decisions will be necessary in order to provide insights pertaining to the rationale the AI model use to draw conclusions and make forecasts. Additionally, frequent trainings and expertise would be needed to increase the interpretability (the propensity of humans to interpret and understand AI algorithms results). The trust of stakeholders on AI based forecast can be increased through robust explainability entailing strong, easy to comprehend and user-friendly justifications on all the decisions and predictions produced by the AI systems. For utilization in grasslands conservation and development, future research need to assess explainable AI types such as (i) opaque systems revealing no reasoning about their algorithmic architecture, (ii) interpretable systems offering mathematical analyses of their algorithmic mechanisms, (iii) systems relying on symbols such as text and visualizations to allow users to understand how algorithm perform conclusions, and (iν) explainable systems that perform automated reasoning to craft concise explanations of their algorithmic mechanisms with minimum human interferences [14].

The assigning of clear responsibility and subsequent accountability of AI based forecasts and decisions continues to remain one of the biggest challenge and same has been realized for their utilization in grasslands conservation. There must be concise responsibility assignment mechanism before operationalization of AI techniques in grasslands in order to avoid potential accidents, and misleading predictions. In addition, there is need to overcome moral proxy problem associated with AI system because these have been designed by humans for serving human ends and to make decisions on their behalf, but these are not advanced enough so far to differentiate morally right or wrong decisions and processes.

Moreover, future research needs to address challenges like AI based model's biasing of results owing to non-representative data, narrow applicability due to non-inclusive usage contexts and their potential failure to differentiate among grass species. In grasslands conservation and development, preventing bias (tendency to learn from a preferred data pattern instead of actual data distribution) and pursuing inclusiveness and fairness in AI based approaches are bound to determine the human-machine partnerships in future. Interestingly, an AI system is generally useful as per assumptions of algorithm, however it can only be expected to precisely forecast decisions based on recorded observations, whereas it make inaccurate predictions in unprecedented, novel and unexpected situations. Likewise, rethinking regarding who will own, store, process and control grasslands data is also a challenge which needs redressal to prevent disputes of ownership, privacy as well as security risks.

7. Conclusions

Owing to climate change, human interferences and ecological pressures, it is high time to implement novel solutions for conserving the natural and semi-natural grasslands. These goals could be effectively achieved through deployment of AI based solution. For instance, use of drones and robots merit to perform conservation actions and reinforce effective management, but multidisciplinary research must resolve the operational and analytical shortcomings that undermine the prospects for their integration in grasslands conservation and development strategies. Future usefulness of AI based solutions will depend on accuracy and reliability of recommendations forecasts made by AI systems. This requires research based development of designs and usability of AI systems involving greater transparency and explainability. Additionally, stakeholders need to focus on governance and outcome based AI approaches involving formal sectors (government or corporate oversight) as well as informal sector (ranchers, tourists etc.). Cantering ranchers and other stakeholders during AI model designing and affixing clear responsibilities along with establishment of a chain of accountability regarding AI decisions may serve as a starting point of AI based solutions for conserving grasslands. Moreover, digitalization of grasslands monitoring operation might increase algorithmic bias risks which necessitate algorithmic objectivity on which AI models thrive and interpret the recorded information. Finally, initially soft compliance must be progressively proceeding to stringent laws for preventing abuse of data and AI systems. However, creative and critical thinking are required to create incentives and establish institutions for ranchers and other human population residing in close vicinity of grasslands to engage them in AI based grassland conservation solutions.

Author details

Muhammad Aamir Iqbal
Faculty of Agriculture, Department of Agronomy, University of Poonch Rawalakot, Azad Jammu & Kashmir, Pakistan

*Address all correspondence to: aamir1801@yahoo.com

References

[1] Iqbal MA. Introductory Chapter: Grasslands Development-Green Ecological Economy and Ecosystem Services Perspectives. London, UK: IntechOpen; 2022. DOI: 10.5772/intechopen.105345

[2] Yu Q, Peng Y, Liu Z, Li T, Hu J, Huang T, et al. Landscape pattern and tourism aesthetic value of shrubs in the mountainous area of West Sichuan, China. Journal of Landscape Research. 2020;**12**(4):71-82

[3] Iqbal MA, Khalid S, Ahmed R, Khan MZ, Rafique N, Ijaz R, et al. Underutilized Grasses Production: New Evolving Perspectives. London, UK: IntechOpen; 2022

[4] Ijaz R, Harun N, Iqbal MA. Grasslannd Development for Ecotourism- Aesthetic Perspectives. London, UK: IntechOpen; 2023

[5] Kujawa S, Niedbała G. Artificial neural networks in agriculture. Agriculture. 2021;**11**(6):497

[6] Jiménez López J, Mulero-Pázmány M. Drones for conservation in protected areas: Present and future. Drones. 2019;**3**(1):10

[7] Sarker IH. Machine learning: Algorithms, real-world applications and research directions. SN Computer Science. 2021;**2**(3):1-21

[8] Benos L, Bechar A, Bochtis D. Safety and ergonomics in human-robot interactive agricultural operations. Biosystems Engineering. 2020;**200**:55-72

[9] Huang CH, Chen PJ, Lin YJ, Chen BW, Zheng JX. A robot-based intelligent management design for agricultural cyber-physical systems. Computers and Electronics in Agriculture. 2021;**181**:105967

[10] Almomani F. Prediction of biogas production from chemically treated co-digested agricultural waste using artificial neural network. Fuel. 2020;**280**:118573

[11] Mohapatra AG, Lenka SK. Neural network pattern classification and weather dependent fuzzy logic model for irrigation control in WSN based precision agriculture. Procedia Computer Science. 2016;**78**:499-506

[12] McCulloch WS, Pitts WH. A logical calculus of the ideas immanent in nervous activity. The Bulletin of Mathematical Biophysics. 1943;**5**:115

[13] LeCun Y, Bengio Y, Hinton G. Deep learning. Nature. 2015;**521**:436-444

[14] Doran D, Schulz S, Besold TR. What does explainable AI really mean? A new conceptualization of perspectives. Available from: https://arxiv.org/abs/1710.00794 [Accessed: January 11, 2024]

[15] Sarker IH, Furhad MH, Nowrozy R. Ai-driven cybersecurity: An overview, security intelligence modeling and research directions. SN Computer Science. 2021;**2**(3):1-18

[16] Koh LP, Wich SA. Dawn of drone ecology: Low-cost autonomous aerial vehicles for conservation. Tropical Conservation Science. 2012;**5**:121-132

[17] Schofield G, Katselidis KA, Lilley MKS, Reina RD, Hays GC. Detecting elusive aspects of wildlife ecology using drones: New insights on

the mating dynamics and operational sex ratios of sea turtles. Functional Ecology. 2017;**38**:42-49

[18] Hassanalian M, Abdelkefi A. Classifications, applications, and design challenges of drones: A review. Progress in Aerospace Science. 2017;**91**:99-131

[19] Sandbrook C. The social implications of using drones for biodiversity conservation. Ambio. 2015;**44**:636-647

[20] Stark DJ, Vaughan IP, Evans LJ, Kler H, Goossens B. Combining drones and satellite tracking as an effective tool for informing policy change in riparian habitats: A proboscis monkey case study. Remote Sensing and Ecological Conservation. 2017;**9**:1-9

[21] Junda J, Greene E, Bird DM. Proper flight technique for using a small rotary-winged drone aircraft to safely, quickly, and accurately survey raptor nests. Journal of Unmanned Vehicle Systems. 2015;**3**:222-236

[22] Zhang J, Hu J, Lian J, Fan Z, Ouyang X, Ye W. Seeing the forest from drones: Testing the potential of lightweight drones as a tool for long-term forest monitoring. Biological Conservation. 2016;**198**:60-69

Chapter 2

Invasive Weeds Dynamics, Plant-Microbes Interactions, and Carbon-Nitrogen Cycles in Sino-Pakistan's Grasslands Perspectives

Chunjia Li, Saima Iqbal, Serap Kizil Aydemir, Xiuqin Lin and Muhammad Aamir Iqbal

Abstract

In China and Pakistan, grasslands serve as carbon sink, ecological barriers, watershed for low riparian regions, feedstock, and minerals extraction sites for drilling and mining and offer numerous associated benefits like wool, herbs for traditional medicines, tourism and leisure, and so forth. However, grassland ecosystems have been persistently degraded by anthropogenic disturbances (land use changes, tourism, intensive grazing, uncontrolled fire, vegetation clearance, invasive weeds, and climate change drivers (heat, drought, chilling, salinity, and shifting of rainfall patterns). To conserve and develop grasslands, soil nitrogen (N) and carbon (C) hold pertinence for maintaining the primary productivity of grass species. Hence, estimating the extent of numerous interventions on N and C cycling along with grass-microbe interactions has become imperative from socioeconomic and environmental perspectives. Thus, to achieve this goal, this chapter has been tailored to compile recent knowledge on the productivity status and persistent degradation of grasslands in China and Pakistan. Additionally, invasive weeds' prevalence in grasslands, grass–microbe interactions and their influence on the growth of plant species, microclimate, and availability of nutrients have been objectively analyzed along with synthesizing the recent advances on C and N dynamics in grasslands ecosystems.

Keywords: grassland conservation, ecosystem development, intensive grazing, parasitic microbes, nitrification, denitrification

1. Introduction

Globally, grasslands (GL) prevail in the regions that generally receive sufficient precipitation to support robust growth of grass species; however, climatic and anthropogenic factors hamper large perennial trees' emergence in these areas. Therefore,

GL occurrence is directly correlated with rainfall (generally in-between of precipitation occurring in deserts and forests) and normally undergoes intensive grazing and wildfire to produce a plagioclimax in previously forested areas. In a narrow sense, GL entail any ground cover having grasses as a predominant vegetation, with meager or no tree cover. Likewise, UNESCO (United Nations Educational, Scientific and Cultural Organization) defines GL as those lands that are dominantly covered by herbaceous plants having tree and shrubs cover of fewer than 10% of the total land area. Interestingly, GL have been classified among world's largest ecosystems by encompassing over 52.5 million km^2 area that accounts for 40% of earth's land surface (excluding Greenland and Antarctica). In contrast, woody savannah occupies 13.8% of the global land area, while shrubs prevail on 12.7% area. Moreover, nonwoody grassland and tundra are situated on 8.3 and 5.7% of earth's land area, respectively [1–3]. The GL offer numerous ecosystem services including habitat provision for thriving flora and fauna biodiversity and contribution to food production systems, along with delivering a wide range of cultural services. The GL store over 34% of carbon stock in the terrestrial ecosystems, of which 90% gets stored belowground as soil organic carbon (SOC), and thus, GL constitute a vital factor in carbon (C) sequestration phenomenon globally. However, GL have remained quite vulnerable to invasive weeds species, anthropogenic disturbances (e.g., land conversion for carrying out modern input-intensive farming, uncontrolled grazing by livestock, etc.), and climate change (CC) [2, 4–6].

In northern China, over 45% of the total GL area is occupied by restored GL, of which over 25% area has undergone serious degradation during last decade. It was estimated that actual net primary productivity (NPP) and human-induced NPP have witnessed a pronounced decrease (0.60 and 5.62 gC m^2 per year, respectively). On the other hand, potential NPP has experienced an increment of 2.27 gC m^2 per year. The CC-associated drivers have been attributed as the prime reasons requiring the restoration of GL situated in the regions of Qinghai, Yunnan, Xinjiang, and Inner Mongolia. Contrarily, anthropogenic interferences were the dominant catalysts for GL degradation, resulting in a reduction of 51932.3 Gg C in their NPP. The human-induced degradation has remained ubiquitous over time in the GL of northeast and northwest China. Except sloppy GL (where climatic change drivers serve as drastic factors), human activities have remained the primary cause of degradation for all types of GL [3, 4, 7–11].

In Pakistan, the predominant GL areas are situated on large swathes of lower Chitral, Waziristan (KPK province), Azad Jammu and Kashmir (AJK), and Gilgit-Baltistan (GB). These GL are mostly found at an altitude of 1400–3600 m, and most of them receive sufficient rainfall (250–750 mm) for robust growth of a wide range of grasses. Like China, GL in Pakistan are also under increasing pressure of anthropogenic perturbations that have noticeably decreased their productivity and ecosystem services. **Figure 1** demonstrates the prominent drivers associated with CC, anthropogenic interferences, and grazing livestock that have induced and triggered land degradation in GL, necessitating their conservation. Fundamentally, the concept of GL conservation entails such practices that are adopted for the protection and sustainable management of GL ecosystems in order to maintain biodiversity and ecological integrity for persistent provision of ecosystem services. In the modern era, GL conservation has become associated with the monitoring of anthropogenic interferences and grass–microbe interactions along with carbon and nitrogen (N) cycles [12–16]. The study of these interactions and nutrient cycles might unveil feasible solution for imparting sustainability to traditional and modern ecosystem services offered

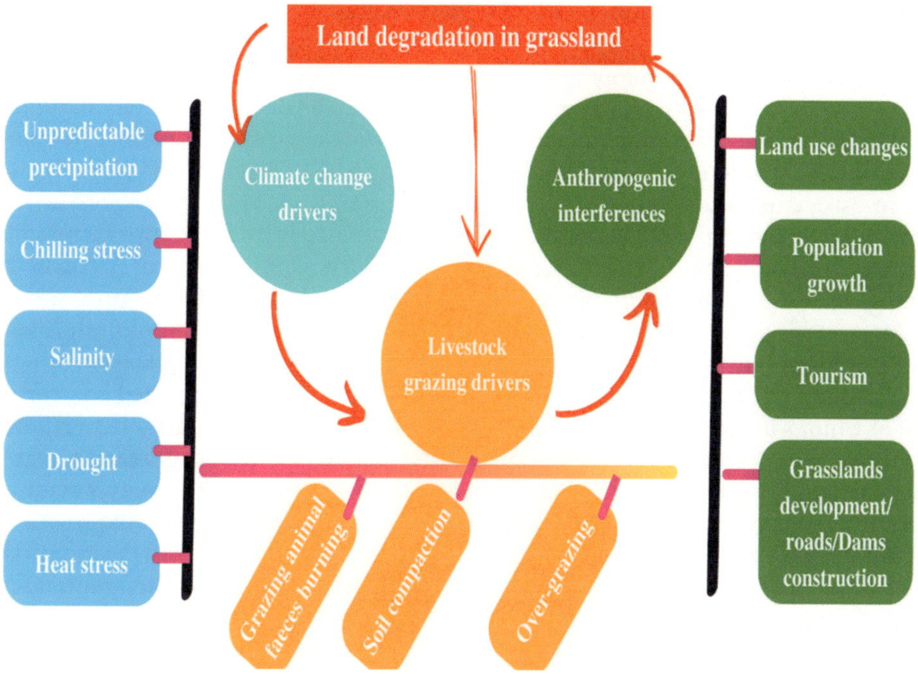

Figure 1.
Pronounced drivers related to anthropogenic interferences (unplanned land use changes, tourism resulting in overutilization of grasslands, and increase in human population directly dependent on grasslands), climate change (salinity, drought, chilling stress, and unpredictable precipitation regimes), and grazing livestock (causing overgrazing without any regard to local vegetation, soil compaction by large hordes, and burning of animal feces) that induce and trigger land degradation in grasslands.

by GL. Moreover, considering the rapidly changing land uses, varying physiographic landscapes, and explicitly diverse ecosystems in the GL of China and Pakistan, it has become increasingly vital to investigate C and N cycles. Furthermore, to impart sustainability to ecosystem services offered by GL biome, it has become crucial to study the C and N alterations and especially the rates of nitrification and denitrification across GL. For instance, studying the implications associated with C and N cycles holds potential to increase our understanding about ecological factors, environmental variables, and human footprint in these vital ecosystems [17–21].

This chapter has been tailored to enhance the existing understanding on the current scenarios of grasslands in China and Pakistan along with synthesizing the recent advances on invasive weeds prevalence (particularly knapweed species, parthenium and Johnsongrass) along with C and N dynamics in grasslands. Three vital questions have been addressed including (i) how do grass–microbe interactions function in GL ecosystems and in what way they effect the growth of plant species, microclimate, and availability of nutrients in the soil solution? (ii) how do different factors regulate C deposition, turnover, and stability in GL ecosystems? and (iii) how do N mineralization, deposition, utilization and losses (as leaching and volatilization) get affected in GL soils? It has been foreseen that assessing the microbe–grass species interaction along with C and N cycling from such perspectives might be crucial to comprehend the impacts of climatic variables and anthropogenic perturbations on GL ecosystems and their implications for the environment in the neighboring countries of Pakistan and China.

2. Current scenario of Sino-Pakistan grasslands

In China and Pakistan, many regions are predominantly natural or improved GL; however, in stricter terms, no GL in these countries can be declared entirely natural. The underlying reason is the involvement of numerous interferences such as human-triggered fires that tend to influence flora and fauna of GL. Another instance of human-associated interference in GL is livestock grazing. Contrastingly, more invasive types of interventions include vegetation clearance to introduce crops farming in the areas of GL and subdivision of GL area with or without temporary or permanent fencing. Moreover, other examples of interferences include water points, provision of grazing area, or season extension along with implementation of numerous improvement techniques (overseeding of leguminous and pasture grasses, application of organic manures and mineral fertilizers, and so on). The GL are expanded on over 40% of China's territory and serve as crucial ecological barrier primarily in northern parts of China. However, these have experienced climate-induced degradation over time, particularly owing to global warming and drought. To make the matter even worse, anthropogenic interferences in the form of land misuse, population growth, and unplanned overgrazing by livestock have further triggered the land degradation in these crucial GL. The net results of these interferences include the emergence of a range of environmental and ecological issues of serious nature over the past couple of decades. Few instances are air and water pollution, natural habitat's unbalancing, and destruction and vegetation degradation. Consequently, land degradation and other disruptions have multiplied economic losses [22–27].

In southwest China, Qinghai–Tibet Plateau (QTP) expands on Tibet and Qinghai province along with large swathes of Sichuan, Yunnan, and Gansu provinces, accounting for around 25% of China's territory. As per estimate, QTP grasslands area represent 30% of China's total GL. The QTP is considered as the highest plateau in the world, with an average elevation of 4000 m [15, 16]. This region is commonly referred to as third pole, roof of the world, hot island, and formation center of numerous species (both flora and fauna). Recently, land degradation progressing to desertification, severe deforestation, and noticeable decline in permafrost have been recorded as the most pronounced cover changes in entire QTP [26–29]. Different types of GL prevalent in QTP include temperate meadow steppe and Alpine meadow steppe, which have been presented in **Figure 2**. Keeping in view the pertinence of GL in northern China in terms of serving as an ecological barrier coupled with a feeding base for livestock production, the initiative involving the installation of enclosures by local governments has emerged as an effective GL restoration measure [30, 31]. However, the impacts of such measures on ecological functioning and production potential of GL still need to be quantified.

Like China, GL in Pakistan are considered very productive in terms of providing numerous ecosystem services. The pastoralist families living around GL get feed for their livestock, and a variety of by-products are prepared from their wool that provide a decent source of income. In true sense, GL-supported animal husbandry holds significant economic pertinence for hundreds of thousands families in Pakistan. The GL biome is situated on large swathes of lower Chitral, Waziristan (KPK province), AJK, and GB, mostly at an altitude of 1400–3600 m with average annual rainfall of 250–750 mm. Most of these GL receive rainfall during the monsoon season. In some regions, most part of precipitation is received in the form of heavy snow especially at higher altitudes of AJK. Overall, dry temperate type of climate having subtropical and subhumid conditions prevails in these GL. Generally, the growing seasons remain

Figure 2.
Different types of grasslands prevalent in QTP regions of China that have been classified on the basis of prevailing climatic conditions, altitude, and dominant type of vegetation.

quite short and cool for most part of the year. The mean temperature ranges from 5–15°C. Among these GL, the Himalaya mountain ranges comprise a wide variety of terrestrial ecosystems having diverse land uses that support human population of over 200 million. The lower Himalaya within Pakistan's geographical boundaries encompass GL, arable, and forest ecosystems that have been subjected to various human perturbations during last couple of decades. Notably, natural forests and GL are being converted into arable areas for shifting subsistence farming to modern high-input commercial farming systems [13–17]. In both Pakistan and China, the grazing systems (GS) in all types of GL are roughly divided into traditional (TGS) and commercial grazing (CGS) systems [32, 33].

3. Invasive weeds in grasslands

Weeds are the unwanted plants that tend to invade, capture the growth resources (moisture, nutrients, solar radiation, space, etc.), and suppress the growth of desired plants. In GL, invasive weeds (IW) reduce biodiversity and degrade their ecosystem functioning to a great extent [34, 35]. Previously, most of the research efforts have been focused on short-term weeds management options, whereas developing means to introduce long-term resistance of GL against IW has been neglected so far. Recently, CC has intensified the problem of IW in grasslands globally and especially in Pakistan and China. These IW reduce the quality of GL along with imparting negative ecological impacts such as biodiversity loss and nutrient cycling alterations [36–38]. A number of factors tend to influence the dynamics of IW invasion such as agro-botanical traits of native grass species, physicochemical characteristics of the soil, and local climate factors. Still, very limited understanding has been unveiled pertaining to the relative importance of environmental conditions on IW invasion dynamics [34, 37].

Recently, in GL of Pakistan, China, USA, and Canada, knapweed species (at least 15 species particularly *Centaurea solstitialis*, *Centaurea diffusa*, *Centaurea maculosa*, *Acroptilon repens,* etc.) primarily originating from Eurasia have emerged as one of the most prime challenges. These species seriously deteriorate the nutritional value of forage grasses, trigger soil erosion, and reduce microbial population in the soil, leading to reduced availability of essential plant nutrients. Besides knapweed species, Johnsongrass (*Sorghum halepense*) has also got the attention of researchers for being one of the most destructive IW in all types of GL [36, 38, 39]. This IW has become prevalent in the GL of Pakistan and China beyond its native origin of Eurasia. Its prevalence is persistently expanding owing to superior botanical traits especially self-pollination-based reproduction, robust growth, wider root networks, and superior adaptability to CC. Resultantly, this IW has seriously reduced native plant diversity in GL on all continents of the globe. Its prevalence has been reported only in tropical and subtropical GL, but it has demonstrated resilience in colder GL of China and Canada as well. Moreover, *Parthenium* (originating from Americas) has now infested GL in over 40 countries of the world [40]. It has invaded GL, and it is causing serious losses to biodiversity, microbial biomass, and deleterious effects for grazing animals and herders. Furthermore, it is increasingly becoming an uncontrolled IW owing to having unique morphological, physiological, and superior ecological adaptive features [40]. Future research is awaited to sort out biologically viable, economically feasible, and environment-friendly solutions to control the population of IW in the GL.

4. Plant–microbe interaction in grassland ecosystem

In all types of GL, soil microbiome–grasses interactions occur in multiple ways that can either boost or hamper the survival and development of one or both of them. Interestingly, numerous biotic and abiotic characteristics of GL's soil are primarily driven by different species of soil microorganisms. Especially, beneficial microbes (commonly referred to as mutualists) and harmful microbes (generally termed as pathogens) can significantly influence the growth and coexistence of grass species with subsequent reciprocal interactions. Recently, this phenomenon has been termed as plant–soil-feedback (PSF). The nature and degree of interaction among individual plant–microbe species has been extensively studied; however, significant gaps exist in our understanding pertaining to the interactive competition and interspecific mutual association involving multiple species of grasses and soil microbes. Soil fertility status imparts strong influence on the type and population growth of different microbes that in result determine the nature of their association with grass species. For instance, pronounced shifts in mycorrhizal colonies occur with soil nutrients availability, while the loads of pathogens tend to multiply with increasing N availability. Alternatively, on a nutrient abundant soil, grass species overcome the negative effects of pathogens by activating their defense mechanisms and strengthening the tolerance level. Thereby, potential adverse effects of pathogens get neutralized and grass plants emerge victorious in their interaction with harmful microbes [41–48].

Another fundamental aspect of grass-microbes' interaction is pertaining to soil organic carbon (SOC) whereby grass species are facilitated by soil microbes to acquire the requisite nutrients. The grass diversity serves as a founding driver for SOC formation and storage in GL soils by elevating belowground C inputs through the addition of root biomass and exudates. Additionally, it also promotes microbial diversity and growth, their turnover, along with necromass entombment. In order to enhance SOC storage in GL's soils, grass biodiversity and C inputs by grass roots are crucial. Moreover, different types of fungi and bacteria tend to impart strong influence on the accumulation, stabilization, and turnover of SOC in GL. Likewise, microbial necromass also triggers SOC accumulation and stabilization with total share of 23–47% in total SOC of GL. Previously, grass–microbial feedbacks have been studied in the competition theory context which suggest that soil microbes tend to shape grass community structure by altering root growth and canopy development. Contrastingly, it has also been indicated that intra- and inter-grass competitions tend to alter the microbial types, their population structure, and growth dynamics of both grass species and soil microbes. However, these findings have limited validity owing to studying grass–microbe interactions on either arable lands or under controlled environments. Overall, interspecific competition among grass species and their interactions with microbial communities simultaneously determine and strongly influence the structure and composition of plant communities in GL. However, accumulating evidences have recently indicated that plant functional groups tend to shape soil and grass roots-linked microbial communities by modifying the soil conditions and quality traits over multiple plant generations [49–54], and this aspect needs further research to explore the dynamics of plant functional groups in GL soils.

The arbuscular mycorrhizal fungi (AMF) residing in grasses' roots tend to derive their C requirement directly from host grass species and also regulate soil C sequestration capacity (CSC). In GL and forests soil, the CSC per unit N may be 1.7-fold greater owing to the domination of ectomycorrhizal fungi–grass species association. The underlying reason is AMF (especially ectomycorrhizal fungi) hold potential to produce enzymes that are vital for degradation of organic N present in grass litter. Contrastingly, many fractions of organic C are present in relatively higher concentrations in ecosystems dominated by AMF especially in GL. It is worth mentioning that climate variables especially temperature and precipitation regulate microbe's metabolic activity and patterns of SOC storage through microbial necromass. In temperate GL, cold and moist soils serve as platform to promote microbial necromass C (MNC) accumulation. The maximum MNC accumulation occurs in GL receiving annual precipitation of over 1000 mm with <0°C mean annual temperature. Thus, microbial diversity present in the root zone determines the SOC storage by regulating MNC and the production of organo-mineral compounds. Furthermore, microbial diversity promotes the grass litter-derived OM stabilization efficiency, which in turn promotes the growth and health of grass species in GL [50, 55–59]. Future research is needed to unveil the impact of climatic factors on mutualistic grass–microbe interactions and decomposition of organic litter for releasing nutrients.

5. Carbon cycle and C-sequestration in grasslands

Globally, GL store over one-third of the terrestrial C stocks and might act as crucial sink of soil C as well. The diversity of grass species increases SOC storage by increasing C inputs to belowground soil horizons and promotes the microbial necromass. The CC has altered the storage capacity of SOC of GL by modifying C inputs processes along with influencing the microbial anabolism and catabolism. The well-planned grazing keeping in view the grass cover of GL and restoration of grass biodiversity may serve as low-cost and effective soil C accumulation to halt the phenomenon of CC globally. It is interesting to note that achievable potentials for SOC sequestration in grasslands for biodiversity restoration and improved grazing management are 2.3–7.3 billion tons of CO_2 equivalents annually and 148–699 megatons of CO_2 year^{-1}, respectively. However, SOC sequestration potential for global GL can be progressively enhanced to 147 megatons of CO_2 year^{-1} if legumes plant population is increased in pasturelands [60–65]. It is high time to initiate overseeding programs for increasing the native legumes percentage in all types of GL.

5.1 Net primary productivity (NPP), actual NPP, and potential NPP

The GL's net primary productivity (NPP) refers to the C quantity fixed by grass species through its utilization in the photosynthesis process within a specific time and particular GL area. The NPP serves as a vital indicator to study C cycle and to assess the growth of diversified vegetation in GL [30]. The NPP of all types of GL in Pakistan and China has remained vulnerable to anthropogenic disturbances and CC; however, it may serve as a valuable tool to enhance our understanding regarding GL response to these stresses. Contrastingly, actual NPP (ANPP) presents insights regarding grass species real conditions and could also be influenced by both meteorological factors as well as anthropogenic drivers. However, potential NPP (PNPP) is solely driven by CC factors and tends to represent the hypothetical growth conditions by excluding the anthropogenic disturbances [11, 56].

5.2 Organic carbon (OC) cycle in grasslands soil

Only 1–2% of OC is present as dissolved organic matter (OM) globally. The OC in the soils of GL is generally distributed into two fractions including particulate organic matter (POM) and mineral-associated organic matter (MAOM). The POM and MAOM are differentiated on the basis of their physicochemical characteristics, mean residing time in soil, and way of formation. The POM formation is primarily regulated by CC drivers especially precipitation and temperature, whereas MAOM formation depends on soil properties (cation exchange capacity, silt and clay content, and availability of microbial released N). Interestingly, the fragmentation of plant and microbial residues results in POM formation; thus, it is generally composed of fragments having lightweight and large polymers. On the other hand, small molecules exuded from the roots of grasses or leaching from grass residues form the MAOM. After microbial assimilation (necromass), these are associated to minerals directly but tend to have significantly lower C:N ratio. The underlying reasons include proportionally greater microbial origin and extended mean residence time in GL soils ranging from decades to centuries. In comparison, POM has residence time of less than 10 years to few decades. It is worth mentioning that 46, 9, and 7% of root exudates, root tissues, and aboveground C residues, respectively, get transformed into MAOM, whereas only a small percentage of root litter (19%) is transformed into POM. Thus, grass species that allocate greater C to roots tend to contribute much higher toward MAOM formation and soil C sequestration in GL. In the top layers of GL soils, 50–75% of SOC is found as MAOM. The average C:N ratios for MAOM and POM vary from 10 to 12 and 16–18, respectively. Therefore, SOC accrual in MAOM requires substantially greater N compared to the equivalent accrual of SOC in POM [22, 56, 57].

5.3 Climate-related drivers and C sequestration in grasslands

Majority (67%) of world's GL are present in arid, semiarid, and cold climates, whereas rest expand into the humid climates. Thus, accrual and stability of C sequestration (CS) in GL is prone to be influenced by CC variables, which tend to exert diverse impacts primarily through grass–microbes mediated mechanisms. These impacts often vary in degree and extent with type of GL and soil along with extent of variation in climate drivers. For instance, higher temperature increases root-derived C input but inhibits the MAOM decomposition through suppression of soil respiration and fungal growth in semiarid steppe. Contrastingly, elevated temperature enhances the vegetation cover of C4 grasses along with triggering the decay rate of SOC fractions in humid tall grass prairies. Likewise, higher temperature in alpine GL causes permafrost degradation, which results in pronounced reduction of SOC storage owing to decreased microbial networks and accelerated decay of SOC. In contrast to MAOM, the POM has remained more sensitive (3-fold higher) to climate variations in different types of GL. It might be inferred that GL having higher MAOM proportion tend to meagerly contribute to carbon-climate feedbacks in GL soil [22, 56].

5.4 Grazing pressure and SOC

Globally, livestock grazing constitutes the most common use of GL. The grazing intensity and regrowth periods often determine the productivity and sustainability of GL under varying temperatures and other climatic variables. There are several drawbacks associated with uncontrolled grazing such as reduction in grasses diversity, cover, and root-microbe-mediated transformation of SOC formation along with increased erosion due to empty patches. Grazing significantly (4–23%) decreases SOC stock as recorded in five continents, whereas 23% reduction in SOC stock was recorded in the tropics, and the corresponding value for temperate GL was 4.5%. Another perspective is natural GL subjected to random grazing tends to have greater SOC storage owing to better grass diversity and cover, mixing of soil layers by trampling of livestock, seed dispersal of deep-rooted grass species, and greater diversity of soil microbes that trigger the formation of OC [64–71].

Additionally, large grazing ruminants work to loosen up soil layers and expose OM's large aggregates to organo-mineral interaction through the promotion of vertical soil mixing. Nevertheless, the extent of impacts imparted by livestock grazing on soil CS is never uniform rather context-dependent in terms of climate variables, soil type, grass species, types of livestock, herbivore diversity, and grazing strategies in terms of grazing intensity and duration. For temperate GL receiving sufficient precipitation, the negative impacts caused by greater grazing intensity are ameliorated, whereas elevated temperatures coupled with extended grazing duration result in significant reduction of SOC. Interestingly, GL entailing C4 grass species record greater SOC in comparison to GL having C3 grasses as dominant vegetation. Moreover, compared to cattle grazing, sheep grazing generally imparts more pronounced negative impacts on SOC especially in the top layers of GL soil. Furthermore, a recent investigation has inferred that soil CS get modulated by soil nutrients availability and grazing intensity of livestock especially in temperate GL [58, 67–69].

5.5 Management options for SOC storage in grasslands

The GL management in a scientific manner holds potential to increase their storage capacity for SOC through mitigation of C losses by climate variables, overgrazing, and other degradation factors [21, 70–74].

1. Compared to crop lands, the GL witness an absence of tillage-related disturbances that results in enhanced root C inputs to GL's soil. For example, grass species diversity restoration triggers soil microbe's turnover and resultant necromass entombment.

2. Additionally, well-managed grazing as per potential of the GL results in remarkably lower C:N ratio, and ultimately, MAOM formation and persistence in soils get skyrocketed.

3. Interestingly, legumes overseeding in GL promotes soil C and N inputs through the enhancement of biomass, excretion of root exudates, and turnover of fine roots.

4. Likewise, manures application (both inorganic and organic ones) holds potential to stimulate net productivity and C inputs to soil leading to efficient microbial C utilization.

5. Well-managed and rationalized grazing (keeping in view the vegetation covers and diversity of grass species), especially by cattle (not sheep or goat), can increase SOC stock by over 28%.

6. Nitrogen cycle in grasslands

Among essential nutrients, N is a vital nutrient for robust growth of grasses, and it is one of the primary limiting nutrients in GL ecosystems [66–70]. Different types of GL produce huge quantities of plant biomass owing to regrowth potential of grass species. The grazing livestock retain only a meager proportion of N contained in forage while its major chunk gets lost through their dung and urine. These grazing animals' excreta add N to the soil as organic N. Thereafter, it undergoes transformation in the soil through the processes of denitrification, leaching, and ammonia volatilization. These transformations result in N losses and ultimately contribute to environmental pollution. During the last couple of decades, significant increment in atmospheric N deposition caused by anthropogenic activities (particularly combustion of fossil fuels) has become the root cause of many ecological disruptions in GL. The atmospheric reactive N gets deposited through wet and dry pathways such as oxidized NOy and reduced NHx. Additionally, N deposition is generally accompanied with the deposition of SO_2^- that depends on different N compounds [26, 59, 61]. **Figure 3** depicts the major components of N along with different sources of N addition and losses from the soil.

6.1 N-uptake, resorption efficiency, and proficiency

1. The N uptake is the process of transporting roots-absorbed N toward the canopy of grasses through xylem. The rate of N uptake depends on external environmen-

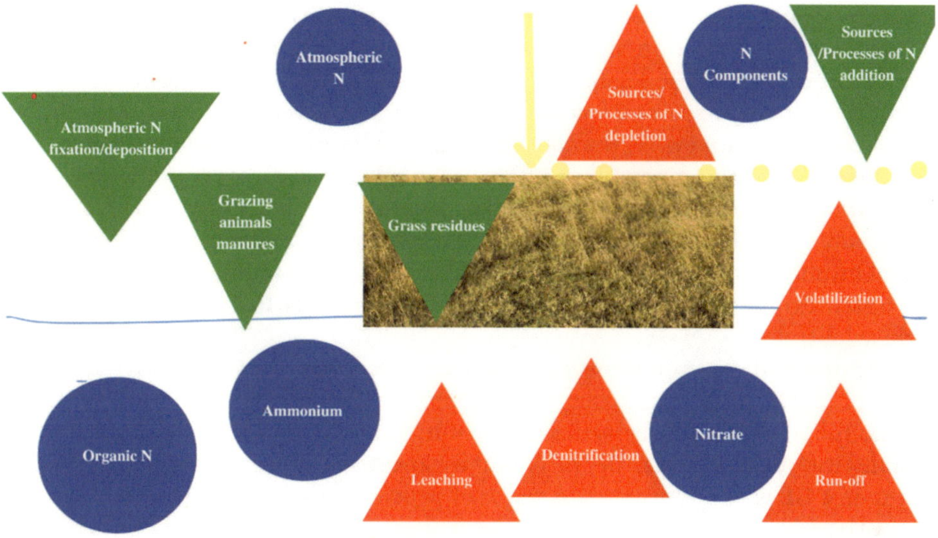

Figure 3.
The nitrogen cycle encompassing vital components or forms of nitrogen in atmosphere and belowground spheres along with processes and sources of nitrogen addition and depletion in a typical grassland ecosystem.

tal drivers, genetic potential of grass species, N concentration in soil solution, and growth stage of grasses.

2. The N resorption refers to the physiological process that entails N transfer from senescent tissues to many other living tissues. This results in increased N residence time within grass plants and enables rapid N recycling between grass plants and the environment. To quantify N resorption, resorption efficiency (NRE) and resorption proficiency (NRP) are generally used.

3. The NRE refers to the percentage of resorbed N in plant tissues and is determined as N concentration of the plant tissue during senescence. It is highly dependent on leaf N status, type of grass species, and GL.

4. The NRP is the degree and extent to which N concentration decreases in the leaf of grasses during aging process over time. For instance, lower N concentration of a senescing leaf exhibits lesser N losses through defoliation and greater N resorption proficiency.

6.2 N dynamics in temperate grasslands of China

In China, temperate GL are expanded on large swathes of land area and offer numerous economic, cultural, and social ecosystem services along with serving as ecological barrier particularly in northern China. Therefore, it is of prime pertinence to investigate the dynamics of N deposition in these GL ecosystems. For instance, in Inner Mongolia's temperate meadow steppe, the rate of N deposition was low; however, it has been advocated as conducive to N addition over time. Additionally, non-N-fixing grass species recorded greater NRE and lower phosphorus resorption efficiency (PRE) compared to N-fixing species. Moreover, N availability in soil solution tends to impart a strong influence on symbiotic N fixation by altering the performance of rhizobacteria in the root zone of plants. Therefore, future research needs to explore the impact of N manures on nutrient resorption and N cycling in GL having sufficient legume cover. The fresh soil collected from GL exhibited greater concentration of ammonia compared to nitrate form of N. However, both were present in noticeably low concentration. Likewise, the concentration of mineralizable N remained in opposition to annual temperature rhythm as winter season recorded the maximum values, whereas summer and early autumn exhibited the minimum values. This might be attributed to addition and/or decay of OM in the soil. The concentration of ammonia and nitrate forms of N was also associated with soil pH as acid soils having pH less than 6 recorded higher concentration of ammonia, whereas less acidic soils chiefly produced nitrate [14, 75–78].

The microbial communities in temperate GL ecosystems chiefly drive and regulate N cycle (addition, losses, and transformation into different N types), whereas inherent edaphic drivers such as inorganic N composition and climate variables influence the quantity of N that escapes into the environment. The N cycling in a GL ecosystem entails a series of processes that work in cohesion to release fixed N into the environment. For instance, nitrification and denitrification may be regarded as the model processes that determine N uptake and losses. It is worth mentioning that nitrification process involves ammonium conversion into nitrates, whereas denitrification refers to the microbial process involving the release of substantial quantities of gaseous N as N_2 and N_2O. These processes are primarily governed by soil microenvironment, whereas microbes play crucial role in N_2O (a potent greenhouse gas) emission. The final products due to incomplete denitrification are N2 and N_2O, which impart pronounced repercussions on the gaseous chemistry of atmosphere as 16% higher N_2O concentrations have been deposited in the atmosphere compared to pre-industrial era. Overall, both of these processes contribute over 70% of the total N_2O emissions from terrestrial ecosystems [14, 72, 79–81]. At local and global levels, mountainous GL serve as a strategic terrestrial ecosystem by providing diverse ecosystem services. However, these vital ecosystems have been persistently under mounting pressure of anthropogenic interferences resulting in distorted N cycling, fixation, and release into the soil and atmosphere. For instance, natural inaccessible forests generally experience meager human interferences and thus record noticeably higher SOM in comparison to heavily grazed GL. Therefore, such scenarios urge future research aimed at monitoring and subsequently formulating effective management plans to ensure sustainable N fixation and lower N emissions from GL ecosystems.

7. Future perspectives of grasslands in Sino-Pakistan context

It has been established beyond any shadow of doubt that GL hold strategic pertinence for both Pakistan and China by offering a range of ecosystem services [1, 3, 12, 13].

1. However, future research especially in temperate GL of Pakistan must tease apart the relative association of grass plant–microbial interactions in terms of mutualism and IW species especially *Parthenium*, knapweeds, Johnsongrass, and so forth. For instance, plant functional groups and allelochemicals secreted by different IW might influence grasses botanical traits particularly root length, which must be studied in conjunction with C and N acquisition strategies as well as microbial taxa present in the root zone of grasses.

2. The extensive GL in Pakistan and northern China must be developed as a strategic feed source for livestock in such a way that ensures sustainable livelihood of stock raisers and herders without compromising the overall health of the vegetation (e.g., allowing grazing in blocks of GL as per capacity of the vegetation cover).

3. Additionally, most of GL in Pakistan and China serve as vital water catchment areas; thus, it has become a matter of prime pertinence to scientifically manage the vegetation cover of these areas in order to ensure sustainability of water supplies for downstream regions.

4. The well-planned management of grazing intensity and maintenance of vegetation cover in GL are bound to reduce the extent of soil erosion and runoff, which seriously damage the water storage infrastructure constructed in the lower catchment areas by causing siltation of reservoirs (Tarbela and Mangla dams in Pakistan are typical instances) and irrigation systems.

5. The scientific-oriented catchment management is bound to incur numerous benefits primarily for communities residing outside of GL premises; however, such maintenance efforts have to be practiced and followed by ranchers in types of GL in Pakistan and China.

6. The GL in northern Pakistan (particularly in Pakistan's administered territory referred to as AJK) and China (Inner Mongolia, Tibet, Sichuan, Yunnan provinces, etc.) are major reserves of biodiversity, providing important wildlife habitat and *in situ* conservation of genetic resources.

7. In some regions of China and Pakistan, GL may be developed for tourism, leisure, and sites of religious significance, which might create additional income for ranchers and communities residing around GL areas, leading to the development of new business arenas.

8. As an effective poverty alleviation measure and to boost the living standards of pastoral communities particularly in GB and AJK regions of Pakistan, the GL must be developed to achieve a wide range of wild foods, herbal medicines, and other useful products such as honey.

9. Being served as large sinks of C, GL might be utilized for recycling of greenhouse gasses and to trigger the buildup of soil OM by restricting tillage practices.

10. The soils of GL receive reasonable quantities of grazing livestock's feces that get deposited as the inorganic N pool, whereas its mineralization enhances associated N losses as leaching and volatilization. Such N losses coupled with added N fertilizers result in the loss of over US$15.9 billion annually [59–61]. Therefore, future research must focus to quantify N addition from animal feces and N losses from GL under CC scenarios. Moreover, grass species preferences for NH_4 or NO_3 might also be studied for estimation of N-cycling pin GL ecosystems.

11. Furthermore, future research needs to reveal the nature of association between the contributions of root exudates, root litter, and other aboveground C inputs to

SOC accumulation (POM and MAOM fraction) with GL types, soil characteristics, and climatic drivers.

12. Keeping in view the pertinence of the subject matter, future research must focus to explore different aspects of N transformation in GL system, N uptake dynamics in grasses and its impact on grasses diversity, N budgeting, and environmental impacts on N manures application in GL.

There are a few other research aspects that need the attention of researchers such as possibility of overseeding C4 grass species along with legumes for improving the N fixation and forage quality of GL [48, 51]. Likewise, future research may unearth the dynamics of negative grass–microbial associations in fertile soils and potential means to shift this interaction into mutualistic association. Similarly, the impacts of climatic variables in conjunction with anthropogenic disturbances on C and N dynamics of different types of GL must be studied. Additionally, key threats such as IW dominating the native grass species, drilling and mining activities, environmental pollution caused by greenhouse gaseous emissions from GL, land degradation, clearing of grasses for converting GL into agricultural lands, overgrazing, bushfires, and so on must be studied to quantify their impact on soil fertility, vegetation productivity, and environmental concerns. Besides, plant–soil feedback has emerged as a vital framework for analyzing multi-trophic interactions among livestock, grass species, and soil microbes residing in the rhizosphere; however, concrete research-based findings are still lacking pertaining to such feedback under CC scenarios. In most of GL, herbivory patterns extensively get affected by CC drivers that further modify the resulting grass–soil feedbacks. Learning from China, there is dire need to develop infrastructure including roads and dams' construction near GL areas in Pakistan keeping in view the geographical dynamics of GL and its ecological consequences.

The GL present in the Hindu-Kush and Himalayan region within Pakistan's geographical boundaries provide a rich resource base for nomadic grazing along with numerous associated ecosystem services. However, GL's quality is seriously decreasing in this region primarily owing to anthropogenic interferences. Therefore, future research planning must focus GL present in high-altitude mountainous areas that are highly susceptible to CC drivers (especially global warming and shift of precipitation patterns). Particularly, the upper Indus basin, a complex area of the Himalayas, has recently witnessed pronounced effects of climatic variability. There is a dire need to formulate policy guidelines with strict implementation plans for the pastoralists in order to ensure sustainable grazing intensity. However, effective monitoring mechanism is a prerequisite to attain reliable information regarding vegetation cover in mountainous GL and impact of CC on soil and vegetation cover [16, 18].

Based on initiatives taken in China regarding the conservation of GL such as construction of fences and boundaries around rapidly degrading GL, it has become evident that until and unless the local dynamics are not appropriately studied and understood, the development of policies and implementation of interventions are less likely to yield desired results. Previously, very scant studies (that too involving only focus group discussions and interviews) have been executed in this region to quantify the impact of CC on vegetation cover in the Himalayas. Therefore, future of these GL lies in conducting a series of systematic assessments to understand GL dynamics in the region, which should formulate the basis for devising future management policies and plans of action. Tackling GL's degradation has remained a daunting challenge in China as well, especially when taking CC into the equation. Future perspectives of northern China's GL lie in expanding the understanding about the drivers that trigger GL degradation especially in Qinghai–Tibet Plateau. Thereafter, bio-economically feasible mitigation measures can be taken to combat this drastic situation for future sustainability. Moreover, such an information might serve as a critical factor for devising conservation and restoration initiatives in different types of GL [31]. The scant understanding of degradation processes and socioeconomic and ecological drivers and absence of long-term monitoring are the real issues that must be redressed for imparting sustainability to ecosystem services offered by GL, and this knowledge gap still needs to be filled within this scope.

8. Conclusion

Grasslands (temperate, alpine, meadow, and steppe) have been confronted with serious challenges by anthropogenic interferences, invasive weed species, environmental variables, and ecological alterations. Future studies need to address emerging challenges associated with vegetation cover contamination by toxic weeds (e.g., *Parthenium*) and anthropogenic-climatic drivers to preserve nitrogen and carbon in GL's soils. Additionally, devising knowledge-based strategies to restore grass species diversity to restrict the invading weeds, preserve SOC and organic N stocks, and promote additional C sequestration for mitigating the climate change phenomenon is the need of the hour. Moreover, future research advances need to highlight the strategic role of soil microbes in regulating the microbial necromass carbon, N transformations, and degradation of organic litter along with C and N storage capacity of GL soils in China and Pakistan. Likewise, the impacts of climate drivers, intensive grazing, wild or well-planned fire, and so forth must be taken into account to devise policies intended for preservation, restoration, and development of GL. Last but not least, future research must be targeted to address the context dependency and uncertainty about the proposed mitigation solutions by considering their trade-offs and possible synergies through collaborative research efforts.

Author details

Chunjia Li[1,2], Saima Iqbal[3], Serap Kizil Aydemir[4], Xiuqin Lin[1,2]
and Muhammad Aamir Iqbal[5*]

1 National Key Laboratory for Biological Breeding of Tropical Crops, Yunnan, China

2 Sugarcane Research Institute, Yunnan Academy of Agricultural Sciences / Yunnan Key Laboratory of Sugarcane Genetic Improvement, Yunnan, China

3 Institute of Microbiology, Government College University, Faisalabad, Pakistan

4 Faculty of Agriculture and Natural Sciences, Department of Field Crops, Bilecik Şeyh Edebali University, Bilecik, Turkey

5 Faculty of Agriculture, Department of Agronomy, University of Poonch Rawalakot, Azad Jammu & Kashmir, Pakistan

*Address all correspondence to: aamir1801@yahoo.com

References

[1] Iqbal MA. Introductory Chapter: Grasslands Development-Green Ecological Economy and Ecosystem Services Perspectives. London, UK: IntechOpen; 2022. DOI: 10.5772/intechopen.105345

[2] Yu Q, Peng Y, Liu Z, Li T, Hu J, Huang T, et al. Landscape pattern and tourism aesthetic value of shrubs in the mountainous area of West Sichuan, China. Journal of Landscape Research. 2020;**12**(4):71-82

[3] Iqbal MA, Khalid S, Ahmed R, Khan MZ, Rafique N, Ijaz R, et al. Underutilized Grasses Production: New Evolving Perspectives. London, UK: IntechOpen; 2022. DOI: 10.5772/intechopen.105375

[4] Li A, Wu J, Zhang X, Xue J, Liu Z, Han X, et al. China's new rural "separating three property rights" land reform results in grassland degradation: Evidence from Inner Mongolia. Land Use Policy. 2018;**71**(2):170-182. DOI: 10.1016/j.landusepol.2017.11.052

[5] Li L, Dong Y, Zhang T, Wang H, Li H, Li A. Environmental and social outcomes of ecotourism in the dry rangelands of China. Journal of Ecotourism. 2022;**22**:1-21

[6] Li WJ, Li YB. Managing rangeland as a complex system: How government interventions decouple social systems from ecological systems. Ecology and Society. 2012;**17**(1):15. DOI: 10.5751/ES-04531-170109

[7] Archer S. Have southern Texas savannas been converted to woodlands in recent history? American Naturalist. 1989;**134**:545-561

[8] Saleh F, Karwacki J. Revisiting the ecotourist: The case of grasslands national park. Journal of Sustainable Tourism. 1996;**4**(2):61-80. DOI: 10.1080/09669589608667259

[9] Long H, Li Y, Liu Y, Woods M, Zou J. Accelerated restructuring in rural China fueled by 'increasing vs. decreasing balance' land-use policy for dealing with hollowed villages. Land Use Policy. 2012;**29**(1):11-22. DOI: 10.1016/j.landusepol.2011.04.003

[10] Wu L, Wang S, Bai X, Tian Y, Luo G, Wang J, et al. Climate change weakens the positive effect of human activities on karst vegetation productivity restoration in southern China. Ecological Indicators. 2020;**115**(1):106392

[11] Shahid N, Yongqiang Z, Jing T, Mueen QF, Aamir L, Kumar PP. Quantifying the impacts of anthropogenic activities and climate variations on vegetation productivity changes in China from 1985 to 2015. Remote Sensing. 2020;**12**(7):1113

[12] Teng M, Zeng L, Hu W, Wang P, Yan Z, He W, et al. The impacts of climate changes and human activities on net primary productivity vary across an ecotone zone in Northwest China. Science of the Total Environment. 2020;**714**:136691

[13] Yang Y, Wang Z, Li J, Gang C, Zhang Y, Zhang Y, et al. Comparative assessment of grassland degradation dynamics in response to climate variation and human activities in China, Mongolia, Pakistan and Uzbekistan from 2000 to 2013. Journal of Arid Environments. 2016;**135**:164-172

[14] Hafeez F, Clément JC, Bernard L, Poly F, Pommier T. Early spring snowmelt and summer droughts

strongly impair the resilience of bacterial community and N cycling functions in a subalpine grassland ecosystem. Oikos. 2023;7:e09836

[15] Maqsood MA, Awan UK, Aziz T, Arshad H, Ashraf N, Ali M. Nitrogen management in calcareous soils: Problems and solutions. Pakistan Journal of Agricultural Sciences. 2016;**53**(1):79-95

[16] Sheikh MA, Anjum J, Tiwari A. Assessment and current dynamics of nutrients with reference to nitrogen attributes in subalpine forests of Western Himalaya. India. Environmental Monitoring and Assessment. 2022;**194**:477

[17] Abbas S, Qamer FM, Murthy MSR, Tripathi NK, Ning W, Sharma E, et al. Grassland growth in response to climate variability in the Upper Indus Basin, Pakistan. Climate. 2015;**3**(3):697-714. DOI: 10.3390/cli3030697

[18] Abbasi AM, Khan MA, Khan N, Shah MH. Ethnobotanical survey of medicinally important wild edible fruits species used by tribal communities of lesser Himalayas-Pakistan. Journal of Ethnopharmacology. 2013;**148**(2):528-536

[19] Begum F, Bajracharya RM, Sharma S, Sitaula BK. Influence of slope aspect on soil physico-chemical and biological properties in the mid hills of Central Nepal. International Journal of Sustainable Development and World Ecology. 2010;**17**(5):438-443

[20] Hosafioğlu I, Akdeniz H, Turan N, Seydoşoğlu S, Iqbal MA, Sabagh AE. Growth, herbage yield and quality evaluation of turfgrasses species under agro-ecological conditions of Iğdir, Turkey. Pakistan Journal of Botany. 2024;**56**(3):1-6. DOI: 10.30848/PJB2024-3(14)

[21] Akdeniz H, Bozkurt MA, Hosaflioglu I, Islam MS, Hossain A, Iqbal MA, et al. Yield and nutritive value of Italian ryegrass (*Lolium italicum* L.) is influenced by different levels of nitrogenous fertilizer. Fresenius Environmental Bulletin. 2019;**28**(12):8986-8992

[22] Dong SK, Zhang J, Li YY, Liu SL, Dong QM, Zhou HK, et al. Effect of grassland degradation on aggregate-associated soil organic carbon of alpine grassland ecosystems in Qinghai-Tibetan plateau. European Journal of Soil Science. 2019;**71**:69-79

[23] Sun J, Cheng GW, Li WP. Meta-analysis of relationships between environmental factors and aboveground biomass in the alpine grassland on the Tibetan plateau. Biogeosciences. 2013;**10**:1707-1715

[24] Wang Z, Zhang Y, Yang Y, Zhou W, Gang C, Zhang Y, et al. Quantitative assess the driving forces on the grassland degradation in the Qinghai–Tibet plateau, in China. Ecological Informatics. 2016;**33**:32-44

[25] Li L, Zhang Y, Wu J, Li S, Zhang B, Zu J, et al. Increasing sensitivity of alpine grasslands to climate variability along an elevational gradient on the Qinghai-Tibet plateau. Science of the Total Environment. 2019;**678**:21-29

[26] Dong J, Cui X, Wang S, Wang F, Pang Z, Xu N. Changes in biomass and quality of alpine steppe in response to N & P Fertilization in the Tibetan plateau. PLoS One. 2016;**11**:e0156146

[27] Wang XY, Yi SH, Wu QB, Yang K, Ding YJ. The role of permafrost and soil water in distribution of alpine grassland and its NDVI dynamics on the Qinghai-Tibetan plateau. Global and Planetary Change. 2016;**147**:40-53

[28] Lu H, Liu G. Trends in temperature and precipitation on the Tibetan plateau, 1961-2005. Climate Research. 2010;**43**:179-190

[29] Liu J, Xu X, Shao Q. The spatial and temporal characteristics of grassland degradation in the three-river headwaters region in Qinghai Province. Acta Geographica Sinica. 2008;**63**:364-377

[30] Wu GL, Du GZ, Liu ZH, Thirgood S. Effect of fencing and grazing on a Kobresia-dominated meadow in the Qinghai-Tibetan plateau. Plant and Soil. 2009;**319**:115-126

[31] Yao X, Wu J, Gong X, Lang X, Wang C, Song S, et al. Effects of long term fencing on biomass, coverage, density, biodiversity and nutritional values of vegetation community in an alpine meadow of the Qinghai-Tibet plateau. Ecological Engineering. 2019;**130**:80-93

[32] Iqbal MA, Iqbal A, Akbar N, Khan HZ, Abbas RN. A study on feed stuffs role in enhancing the productivity of milch animals in Pakistan-existing scenario and future prospect. Global Veterinaria. 2014;**14**(1):23-33

[33] Iqbal MA. Evaluation of forage cowpea and hey as a feed resource for ruminant production: A mini-review. Global Veterinaria. 2014;**14**(5):747-751

[34] Alday JG, Cox ES, Pakeman RJ, Harris MPK, Le Duc MG, Marrs RH. Effectiveness of Calluna-heathland restoration methods after invasive plant control. Ecological Engineering. 2013;**54**:218-226. DOI: 10.1016/j.ecoleng.2013.01.038

[35] Goslee SC, Peters DPC, Beck KG. Modeling invasive weeds in grasslands: The role of allelopathy in Acroptilon repens invasion. Ecological Modelling. 2001;**139**:31-45. DOI: 10.1016/S0304-3800(01)00231-9

[36] Gaskin JF, Espeland E, Johnson CD, Larson DL, Mangold JM, et al. Managing invasive plants on Great Plains grasslands: A discussion of current challenges. Rangeland Ecology & Management. 2021;**78**:235-249. DOI: 10.1016/j.rama.2020.04.003

[37] Bakker JD, Wilson SD, Christian JM, Li X, Ambrose LG, Waddington J. Contingency of grassland restoration on year, site, and competition from introduced grasses. Ecological Applications. 2003;**13**(1):137-153. DOI: 10.1890/1051-0761(2003)013[0137:COGROY]2.0.CO;2

[38] Kim KD, Ewing K, Giblin DE. Controlling Phalaris arundinacea (reed canarygrass) with live willow stakes: A density-dependent response. Ecological Engineering. 2006;**27**(3):219-227. DOI: 10.1016/j.ecoleng.2006.02.007

[39] Mahaney WM, Gross KL, Blackwood CB, Smemo KA. Impacts of prairie grass species restoration on plant community invasibility and soil processes in abandoned agricultural fields. Applied Vegetation Science. 2015;**18**(1):99-109. DOI: 10.1111/avsc.12128

[40] Belgeri A, Bajwa AA, Shabbir A, Navie S, Vivian-Smith G, Adkins S. Managing an invasive weed species, Parthenium hysterophorus, with suppressive plant species in Australian grasslands. Plants. 2020;**9**(11):1587. DOI: 10.3390/plants9111587

[41] Kardol P. Microbe-mediated plant–soil feedback causes historical contingency effects in plant community assembly. Ecological Monographs. 2007;**77**:147-162

[42] Reynolds HL. Grassroots ecology: Plant–microbe–soil interactions as

drivers of plant community structure and dynamics. Ecology. 2003;**84**:2281-2291

[43] Mariotte P. Plant–soil feedback: Bridging natural and agricultural sciences. Trends in Ecology & Evolution. 2018;**33**:129-142

[44] Vincenot CE. Plant–soil negative feedback explains vegetation dynamics and patterns at multiple scales. Oikos. 2017;**126**:1319-1328

[45] Van Nuland ME. Plant–soil feedbacks: Connecting ecosystem ecology and evolution. Functional Ecology. 2016;**30**:1032-1042

[46] Gustafson DJ, Casper BB. Nutrient addition affects AM fungal performance and expression of plant/fungal feed-back in three serpentine grasses. Plant and Soil. 2004;**259**:9-17

[47] Wilson GWT, Hartnett DC. Interspecific variation in plant responses to mycorrhizal colonization in tallgrass prairie. American Journal of Botany. 1998;**85**:1732-1738

[48] Jiang J. Plant–mycorrhizal interactions mediate plant community coexistence by altering resource demand. Ecology. 2017;**98**:187-197

[49] Whitaker BK et al. Viral pathogen production in a wild grass host driven by host growth and soil nitrogen. The New Phytologist. 2015;**207**:760-768

[50] Yang GW et al. The interaction between arbuscular mycorrhizal fungi and soil phosphorus availability influences plant community productivity and ecosystem stability. Journal of Ecology. 2014;**102**:1072-1082

[51] Day NJ et al. Temporal dynamics of plant–soil feedback and root-associated fungal communities over 100 years of invasion by a non-native plant. Journal of Ecology. 2015;**103**:1557-1569

[52] De Deyn GB, van der Putten WH. Linking above-ground and belowground diversity. Trends in Ecology & Evolution. 2005;**20**:625-633

[53] Frouz J et al. Effects of soil substrate quality, microbial diversity and community composition on the plant community during primary succession. Soil Biology and Biochemistry. 2016;**99**:75-84

[54] Garcia-Parisi PA, Omacini M. Arbuscular mycorrhizal fungi can shift plant–soil feedback of grass–endophyte symbiosis from negative to positive. Plant and Soil. 2017;**419**:13-23

[55] Harpole WS, Suding KN. A test of the niche dimension hypothesis in an arid annual grassland. Oecologia. 2011;**166**:197-205

[56] Harpole WS et al. Addition of multiple limiting resources reduces grassland diversity. Nature. 2016;**537**:93-96

[57] Hartnett DC, Wilson GWT. Mycorrhizae influence plant community structure and diversity in tallgrass prairie. Ecology. 1999;**80**:1187-1195

[58] Hoeksema JD et al. A meta-analysis of context-dependency in plant response to inoculation with mycorrhizal fungi. Ecology Letters. 2010;**13**:394-407

[59] Johnson NC et al. Functioning of mycorrhizal associations along the mutualism–parasitism continuum. The New Phytologist. 1997;**135**:575-585

[60] Gao T, Xu B, Yang X, Jin Y, Ma H, Li J. Review of researches on biomass carbon stock in grassland ecosystem of Qinghai-Tibetan plateau. Progress in Physical Geography. 2012;**31**:1724-1731

[61] Yu L, Chen Y, Sun W, Huang Y. Effects of grazing exclusion on soil carbon dynamics in alpine grasslands of the Tibetan plateau. Geoderma. 2019;**353**:133-143

[62] Lu F, Hu H, Sun W, Zhu J, Liu G, Zhou W, et al. Effects of national ecological restoration projects on carbon sequestration in China from 2001 to 2010. The Proceedings of the National Academy of Sciences. 2018;**115**:4039-4044

[63] Wang XX, Dong SK, Yang B, Li YY, Su XK. The effects of grassland degradation on plant diversity, primary productivity, and soil fertility in the alpine region of Asia's headwaters. Environmental Monitoring and Assessment. 2014;**186**:6903-6917

[64] Lu X, Kelsey KC, Yan Y, Sun J, Wang X, Cheng G, et al. Effects of grazing on ecosystem structure and function of alpine grasslands in Qinghai–Tibetan plateau: A synthesis. Ecosphere. 2017;**8**:e01656

[65] Li Y, Li J, Are KS, Huang Z, Yu H, Zhang Q. Livestock grazing significantly accelerates soil erosion more than climate change in Qinghai-Tibet plateau: Evidenced from 137Cs and 210Pbex measurements. Agriculture, Ecosystems and Environment. 2019;**285**:106643

[66] Sun J, Hou G, Liu M, Fu G, Zhan T, Zhou H, et al. Effects of climatic and grazing changes on desertification of alpine grasslands, Northern Tibet. Ecological Indicators. 2019;**107**:105647

[67] Harris RB, Samberg LH, Yeh ET, Smith AT, Wang WY, Wang JB. Rangeland responses to pastoralists, grazing management on a tibetan steppe grassland, Qinghai province, China. Rangeland Journal. 2016;**38**:1-15

[68] Dong SK, Shang ZH, Gao JX, Boone R. Enhancing sustainability of grassland ecosystems through ecological restoration and grazing management in an era of climate change on Qinghai-Tibetan plateau. Agriculture, Ecosystems and Environment. 2020;**287**:106684

[69] Liu S, Zamanianb K, Schleussa P, Zarebanadkoukic M, Kuzyakova Y. Degradation of Tibetan grasslands: Consequences for carbon and nutrient cycles. Agriculture, Ecosystems and Environment. 2018;**252**:93-104

[70] Su XK, Wu Y, Dong SK, Wen L, Li YY, Wang XX. Effects of grassland degradation and re-vegetation on carbon and nitrogen storage in the soils of the headwater area nature reserve on the Qinghai-Tibetan plateau, China. Journal of Mountain Science. 2015;**12**:582-591

[71] Wen L, Jinlan W, Xiaojiao Z, Shangli S, Wenxia C. Effect of degradation and rebuilding of artificial grasslands on soil respiration and carbon and nitrogen pools on an alpine meadow of the Qinghai-Tibetan plateau. Ecological Engineering. 2018;**111**:134-142

[72] Sadiq M, Rahim N, Iqbal MA, Alqahtani MD, Tahir MM, Majeed A, et al. Rhizobia inoculation supplemented with nitrogen fertilization enhances root nodulation, productivity, and nitrogen dynamics in soil and black gram (Vigna mungo (L.) Hepper). Land. 2023;**12**:1434

[73] Kizilgeci F, Yildirim M, Islam MS, Ratnasekera D, Iqbal MA, Sabagh AE. Normalized difference vegetation index and chlorophyll content for precision nitrogen management in durum wheat cultivars under semi-arid conditions. Sustainability. 2021;**13**:3725

[74] Yildirim M, Kizilgeci F, Albayrak O, Iqbal MA, Akinci C. Grain yield and nitrogen use efficiency in spring wheat

(Triticum aestivum L.) hybrids under different nitrogen fertilization regimes. Journal of Elementology. 2022;**27**(3):627-644

[75] Ali S, Fatima A, Bukhari QA, Iqbal MA, Rana MA, Afzaal S. Effect of nitrogen fertilization on polyphenols in Trachyspermum ammi and Foeniculum vulgare under various lights interception. Z Arznei-Gewurzpfla. 2019;**23**(2):67-71

[76] Chen D, Huang H, Hu M, Dahlgren RA. Influence of lag effect, soil release, and climate change on watershed anthropogenic nitrogen inputs and riverine export dynamics. Environmental Science & Technology. 2014;**48**(10):5683-5690

[77] Di HJ, Cameron KC, Podolyan A, Robinson A. Effect of soil moisture status and a nitrification inhibitor, dicyandiamide, on ammonia oxidizer and denitrifier growth and nitrous oxide emissions in a grassland soil. Soil Biology and Biochemistry. 2014;**73**:59-68

[78] Legay N, Grassein F, Robson TM, Personeni E, Bataillé MP, Lavorel S, et al. Comparison of inorganic nitrogen uptake dynamics following snow melt and at peak biomass in subalpine grasslands. Biogeosciences. 2013;**10**:7631-7645

[79] Hayakawa A, Nakata M, Jiang R, Kuramochi K, Hatano R. Spatial variation of denitrification potential of grassland, windbreak forest, and riparian forest soils in an agricultural catchment in eastern Hokkaido, Japan. Ecological Engineering. 2012;**47**:92-100

[80] Sgouridis F, Ullah S. Denitrification potential of organic, forest and grassland soils in the Ribble-Wyre and Conwy River catchments, UK. Environmental Science: Processes & Impacts. 2014;**16**(7):1551-1562

[81] Pommier T, Cantarel AAM, Grigulis K, Lavorel S, Legay N, Baxendale C, et al. The added value of including key microbial traits to determine nitrogen-related ecosystem services in managed grasslands. Journal of Applied Ecology. 2018;**55**:49-58

Chapter 3

Grasslands Development for Ecotourism: Aesthetic Perspectives

Raina Ijaz, Nidaa Harun and Muhammad Aamir Iqbal

Abstract

Grasslands (also known as savanna, prairie, steppe, and pampas) are natural or seminatural areas encompassing vegetation belonging to the family Poaceae as the most dominant vegetation, while, sedges and rushes may also constitute a minor proportion. These provide numerous natural products such as food feed medicinal raw material, and honey along with nonproduct-based ecosystem services. Grasslands in lowlands and mountains either in natural form or developed landscape can provide an added value in terms of ecotourism opportunities owing to having huge esthetic and recreational potential compared to uniform agricultural areas. Grasslands characterized by high species and habitat diversity-based ecotourism are nature-based tourism whereby people visit natural or developed areas for recreation, sight-seeing, permitted and controlled hunting, on-site purchase of organic products, etc., and are usually managed by adopting sustainable practices. Ecotourism generates multifaceted economic advantages for local communities such as direct sale of products to tourists. However, ecotourism may also have a variety of negative impacts when the tourists' number multiplies which leads to overuse of resources. The most pronounced challenges confronted to the development of grasslands for ecotourism include lack of community cooperation, careless herders, need of hefty investment, and absence of trained human capital along with climate change and loss of biodiversity.

Keywords: agronomy and horticulture, prairie, pampas, herders, rangelands

1. Introduction

Grasslands (natural or developed areas entailing predominantly grass species with meager shrubs along with other vegetation like trees) are one of the vital elements of terrestrial ecosystem for having economic, social, ecological, and cultural roles [1, 2]. The raising of dairy animals in a grassland-supported feeding system provides farmers with decent revenues. These not only constitute as the vital source of feed for milch and draught animals but also their importance in terms of forage quality association with animal-origin products is beyond any shadow of doubt [3, 4]. Grasslands hold potential to not only provide hefty quantities of food and fiber but also play a pertinent role in imparting sustainability to the ecosystem. It deserves mentioning that grassland ecosystems represent a vital sustainer of biodiversity by offering optimal conditions pertaining to diversified species and habitats for birds, animals, and invertebrates. However, in order to halt biodiversity loses, there is dire need to strive for conserving the grasslands through wise and sustainable uses. Along with habitat protection and species conservation, grasslands contribute to enhancing the quantity and value of traditional products (medicinal raw material, forage, grasses for beverage, honey, etc.) along with non-commodity outputs in the form of balanced ecosystem functioning and resilience to environmental paradigms [5, 6]. Different types of grasslands (natural, man-made, developed, seminatural, tropical, temperate, etc.) tend to produce different quantities and qualities of ecosystem services (ES), while a variety of factors such as vegetation cover, climate, and geography are taken into account for classifying the grasslands as depicted in **Figure 1**.

Globally, the pertinence of grasslands becomes even more important as over one-quarter of the earth's land surface is covered with grassland ecosystems [1, 2]. There exist vast grassland systems in the planes of North America, sub-Saharan Africa, South America, Australia, and Asia. These are the chief source of dairy products and meat, along with accounting for about one-third of the total carbon of all terrestrial ecosystems. These provide numerous ES such as night cooling, soil erosion control, and flood mitigation [7, 8]. As per agricultural perspectives, there are three major types of grasslands including natural, seminatural, and improved grasslands [9, 10]. Natural grasslands basically form the grassland biomes and these were created by practices related to climate, fire, and wildlife grazing [11], but are also used by livestock. In comparison to natural, seminatural grasslands are formed through human management, and for their maintenance, it needs livestock grazing or hay-cutting [12]. Whereas, improved grasslands are pastures that are formed by plowing and sowing agricultural varieties or non-native grasses having high production potential. Artificial fertilizers and intensive management are required for maintenance [13, 14]. The former two grassland types need conservation attention because of their biodiversity significance, reduction in area, and the fact that their full capacity to deliver multiple types of ES. **Figure 2** presents different ecosystems services offered by the grasslands to improve ecosystem functioning and diversified utilities for humans.

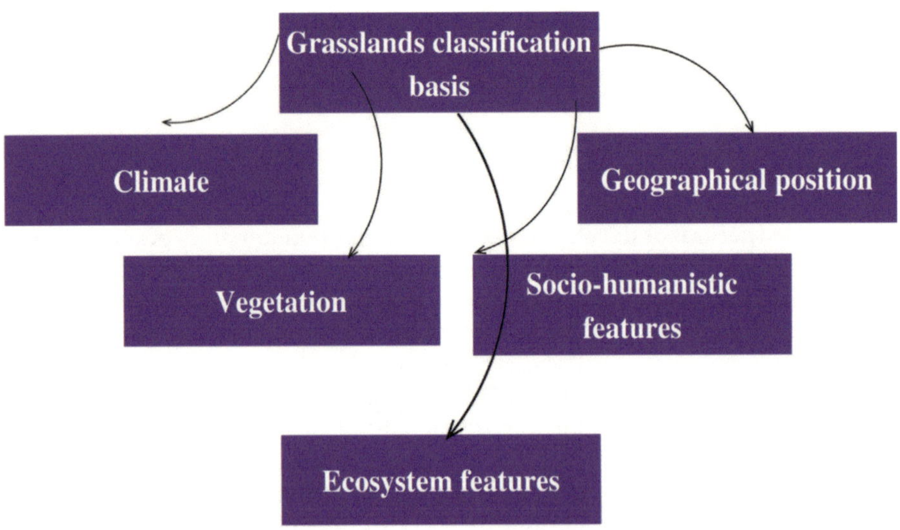

Figure 1.
Different factors for classifying the grasslands such as climate, geography, and socio-humanistic.

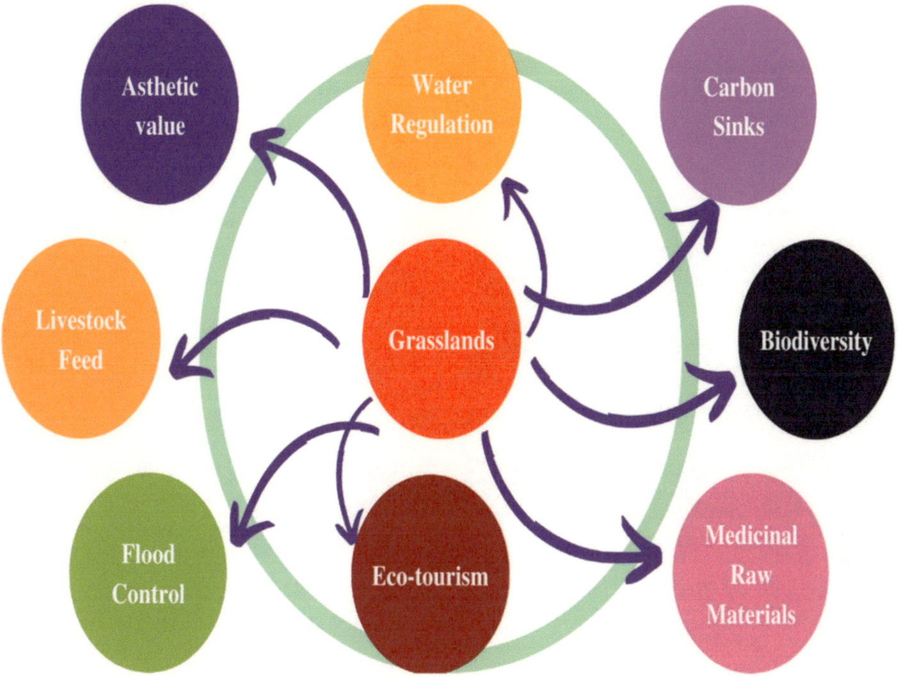

Figure 2.
Different ecosystem services offered by the grasslands to improve the ecosystem functioning and other diversified utilities for humans.

2. Esthetic value of grasslands

One of the primary reasons grasslands hold such vibrant appeal is their inherent esthetic beauty. The visual appeal of grassland landscapes lies in their simplicity and vastness. The open expanse of grasses swaying gently in the wind, the vibrant colors of wildflowers scattered across the fields, and the panoramic views stretching to the horizon create a sense of tranquility and awe. The changing colors of greenery, diversity of wildlife in natural or developed grasslands, and chirping of different species of birds create amazing scenes and enhance the esthetic value of grasslands. Likewise, the esthetic perspective of grasslands becomes undeniable owing to unique beauty of this type of ecosystems that appeals to tourist's senses. Additionally, the visual attraction of grasslands lies in their open spaces and attention-grabbing colors of different types of vegetation. Moreover, the tourists are attracted to the calmness of environment in grasslands whereby grasses movement with parcel of wind creates a unique landscape.

Grasslands also possess unique features and natural attractions that contribute to and enhance their unmatched ornamental value. These include striking geological formations, such as limestone outcrops or rocky ridges, which add texture and character to the landscape. The presence of rivers, wetlands, and lakes within grassland ecosystems adds further diversity and visual interest. These features provide opportunities for eco-tourists to engage in activities like bird watching, hiking, or wildlife spotting, enhancing the overall experience of the grassland environment. Esthetic experience reflects the intimate relation of people with their ecological system [15]. This experience fluctuates as per levels of environmental organization and level of

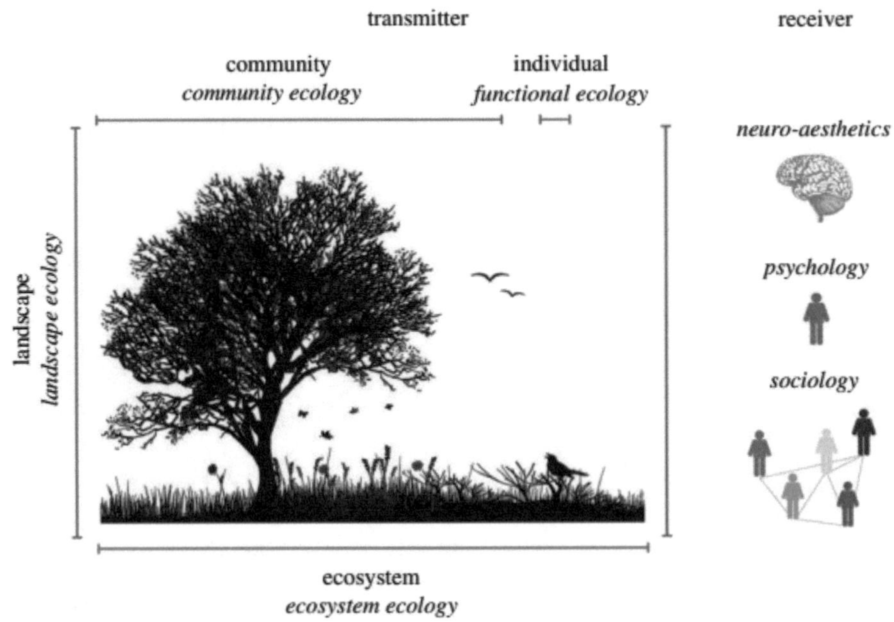

Figure 3.
The levels of esthetic perception in the perspectives of ecosystem ecology.

human perception (**Figure 3**). Esthetic value strongly influences people's inspiration for biodiversity conservation at the landscape as well as species levels [15–17]. Esthetic perception of humans varies according to their level of organization in ecosystem and scale at which humans integrate this information. Functional ecology, community and ecosystem ecology, and landscape ecology are important to study different organization levels of transmitter. Neuroaesthetics and psychology help to study the cognitive processes correlating visual information to emotion from the receiver point of view. Influence of landscape perception on human behavior or mental health can be studied through social science and psychology. The relationship between culture and nature can be studied via fields of philosophy, art, and humanities (e.g., symbolic values and sense of place) [18, 19].

3. Threats to grasslands ecosystem

From past century, grassland ecosystems are facing declining around the world and this decline is still continuing. Both natural disturbances (biotic and abiotic stresses) and anthropogenic influences are the causes for this grassland degeneration [20, 21]. Agricultural activity can be considered as the one of major threat to these systems as occurring on 7.1 million km^2 of the earth's total grassland biome. Reduction in species diversity is also consequence of loss of grassland habitats because of habitat fragmentation, over exploitation by local grazers, amplified nitrogen deposition from the atmosphere, and alterations in fire frequency. Impacts of human activities not only resulted in biodiversity loss at local but also at global level [22, 23]. **Figure 4** presents pronounced threats confronting to the ecosystems of the grasslands.

The survival of remaining biodiversity will only be possible if humans willing to allocate their services and funds for its conservation. Hence, it can be said

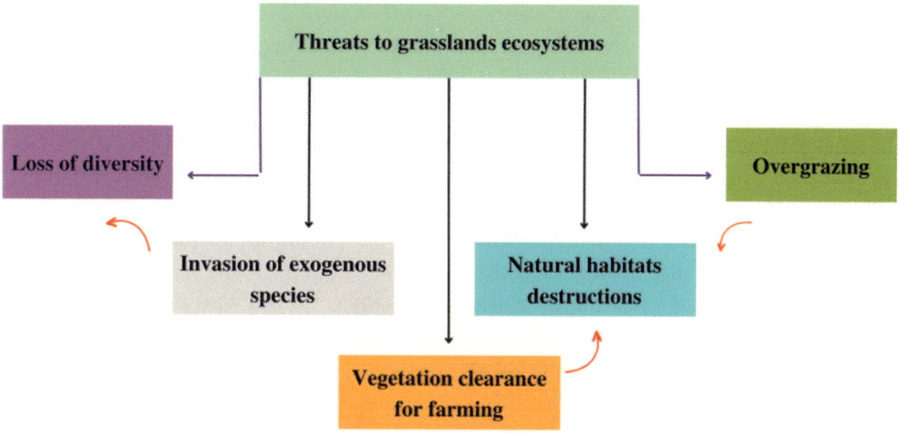

Figure 4.
Pronounced threats confronting to the ecosystems of the grasslands.

that biodiversity conservation critically depends on the basic human concerns for ecological, economic, esthetic, cultural, ethical, and spiritual. These above-stated values motivate communities to support conservation. Nevertheless, human biodiversity preferences are poorly understood by conservationists. Interpretations about public preferences might allow even better prediction of the acceptance of biodiversity management actions, and facilitate the development of appropriate ways of collaborating these, hence, increasing the probability of biodiversity management success [3, 24].

These above-stated ecological issues with the emerging trends of tourism reinforced the development of ecotourism in natural territories. The development of ecotourism activities will inevitably impart positive impacts on the living environment of wild flora and fauna in protected areas. Ecotourism holds the potential to be developed as a promising strategy to channelize tourism-based revenues for promoting environmental conservation and contributing to poverty alleviation. However, vulnerable social-ecological conditions may limit the effects of ecotourism in dry rangelands around the world. However, detailed and target-oriented research must be conducted to promote ecotourism construction by maintaining a balance between conservation and livelihood of local communities [25]. To ensure the conservation and development of grasslands ecosystem having robust ecological diversity, over-seeding of ornamental grasses in natural grasslands might be developed as a critical strategy to promote ecotourism. Supplementation of ornamental grasses with lawn plantation has valuable impact in terms of esthetic, recreational, and health-hygiene.

4. What is ecotourism?

In terms of defining ecotourism, it must be stated that it continues to remain a debatable topic so far among researchers and policy makers. The International Union for Conservation of Nature elaborates ecotourism as an environmentally responsible traveling to an undisturbed natural area with an aim to study, enjoy, and/or appreciate nature which leads to the promotion of conservation of that natural area along with having meager negative visitor's impact, and offers active venues of socioeconomic involvement for the local community. Another definition of ecotourism entails that it is a business that organizes holidays to different places of natural beauty in such a manner that assists local people economically without damaging the environment. Likewise, Global Ecotourism Network defines ecotourism as the responsible travel to natural areas in such a way that conserves the environment, sustains the local people socially and economically along with creating the knowledge and understanding by educating all stakeholders. Another definition of ecotourism encompasses it as an activity for experiencing and learning about natural areas, their landscape (flora, fauna, and their habitats) and cultural artifacts from the surrounding locality. It further illustrates that a symbiotic relationship between the tourist activities and environment is developed when ecotourism theories are translated into appropriate policy initiatives, carefully articulated planning, and tactful practicum. Ecotourism has been regarded as a specific form of nature-based tourism or a nature-based traveling in the field of tourism. The most pronounced characteristic of ecotourism is that it is nature-based tour or travel for recreation without damaging the environment. Another definition of ecotourism regards it as a form of tourism that is inspired primarily by the history of a natural area and its indigenous cultural and traditional values. The ecotourists usually visit relatively undeveloped regions in the spirit of participation, appreciation, and sharing knowledge and cultural exchange with natives of the area. The ecotourists practice a nonconsumptive use of natural resources and indigenous wildlife which contributes to the site's conservation and economic upliftment of the local communities [26–28].

Theoretically, ecotourism has emerged as one of the most sustainable, economically viable, pro-environment, and promising solutions to attain the goals of zero hunger, poverty alleviation, conservation, and local development. The underlying factor is ecotourism can effectively channelize revenues generated from tourism towards the support services aimed at conserving the environment. In addition, it holds potential to serve as an alternative income source for local people and resultantly they are enabled to reduce their dependency exploitation grassland's ecosystems and local wildlife including highly endangered species that are at the brink of complete extinction. Moreover, it assists to transmit traditional ecological knowledge in a systematic way along with improving public awareness regarding environmental crises, especially in far-flung and marginal regions of the globe. Therefore, ecotourism especially community-based ecotourism needs to be developed as a critical path for attaining the UN Sustainable Development Goals [29–35].

Most prominently, community-based grasslands ecotourism can effectively help to achieve SDG-1 (No poverty), SDG-8 (decent work and economic growth), SDG-14 (Life below water), and SDG-15 (Life on land). However, this potential of grasslands ecotourism has still not been realized even partially owing to a variety of hurdles and challenges. The most pronounced limiting factors include the shortages of requisite financial resources, trained human capital, and social capital having low level of awareness pertaining to the potential of grasslands ecotourism. These limiting factors are further worsened by the imperfection of tourism market, especially the conspicuous consumption and green washing business. Another vital aspect as in many cases, grassland ecotourism fails to aid community development and grasslands conservation, which has raised serious questions on the idealistic claims made by many sociologists, economists, researchers, and scholars regarding the potential of grassland ecotourism [36–39].

Interestingly, policy makers are confronted to a bigger challenge pertaining to the development of ecotourism in dry land's grasslands, which occupy over 28% of the globe's land area and support around 2 billion people in direct or indirect way. These grasslands effective management, conservation, and development have become critical in order to achieve global sustainability. **Figure 5** presents different vital elements of an effective ecotourism policy. Grasslands and community-based ecotourism have successfully played a strategically vital role in the local economies of many developed countries of North America, Australia, and Europe by achieving a sustainable balance of trade-offs between grasslands conservation and development and means of livelihoods. However, it deserves mentioning that in developing countries of South Asia, especially the Indo-Pak subcontinent, most of the herders have meager access to the markets, public services, technical knowledge, infrastructure, etc., which has prevented the development of grassland ecotourism [10, 40]. Likewise, a wide range of grasslands ecotourism-associated undesirable outcomes have emerged in China, including fraud, exorbitant charges, severe damage by off-road vehicles, and more pronouncedly, eruption of conflicts between local herders and tourists. These undesirable results indicate serious knowledge gaps in applying ecotourism theories without giving due consideration to local economic needs and cultural values.

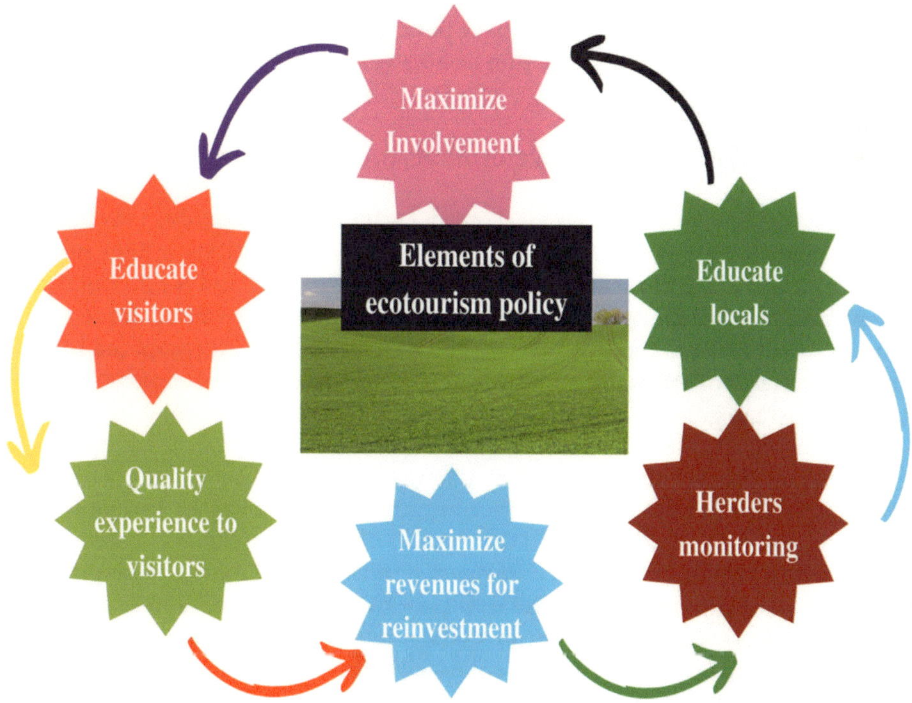

Figure 5.
Elements of an effective ecotourism policy covering aspects related to local community, grassland conservation, and tourists' perspectives.

5. Case studies

5.1 Ecotourism in Ergun grassland

Socioecological effects of ecotourism were surveyed in one of China's most-admired ecotourism regions, that is, Ergun grassland. In comparison to livestock feeding, ecotourism at local helped to attain multiple sustainable goals, like source of income for natives, development of community cooperation, and awareness about conservation of local natural resources. Nevertheless, this affects the diversity of forb species and succeeding reduction in ES. Hence, it is considered that ecotourism in the particular region, still requires improvement and only extensive research which offers economically and biologically viable solutions to local challenges can promote ecotourism in natural, seminatural, or improved grasslands [41].

5.2 Kalahari's landscape

Botswana has various kinds of ecosystems which are enriched with diversified wild species. Kalahari is one the important semidesert of the country that covers the 84% area of whole country. The landscape of Kalahari is dominant with grasslands, scattered trees, and xerophytic vegetation. The Okavango Delta, the Savuti, and the Chobe are at the northeastern sides of Kalahari. These are much wetter areas and rich with diverse wildlife species which are supported by grassland ecosystem. Botswana is unique in that most of its biodiversity is conserved, with a higher percentage of its total landmass conserved than any other country. This conservation level is chiefly achieved through ecotourism. Government policy, high-income, and low-volume tourism support ecosystem conservation in Botswana [42–44].

6. Advantages of grasslands development for ecotourism

Globally, intensively managed grasslands have long been recognized for being the site of conservation for plant and animal biodiversity along with their huge potential for social and cultural utilizations. For the time being, natural and seminatural grassland's capacity to deliver a variety of ES as a part of modern and commercially oriented agricultural systems has surprisingly been understudied compared to other natural resources. Theoretically, it may be perceived that in case of income generation from ecotourism in grasslands, locals might be pursued to exploit lesser biomass from the grasslands. In this way, the conserved biomass may be allocated to support the regulating and other cultural services providers. Thus, development of grasslands for ecotourism holds potential to serve as an alternative and sustainable income source for locals. In addition, it can suppress the chances of conflicts eruption between conservation efforts and livelihood. However, it must be kept in mind that grasslands ecotourism can be promoted at much faster pace by following the community and culturally based principles and traditions. The underlying reason is community-based ecotourism tends to fairly share the revenues with local people, which in turn might assists in developing local institutions for sustainable management of grasslands and wildlife in accordance with devised policies and practices. **Figure 6** illustrates numerous pronounced advantages that can be extracted by promoting grasslands as an avenue of ecotourism. To sum up, grasslands ecotourism must be considered as for, by, and with local community, which in long run, can assist to achieve the sustainable

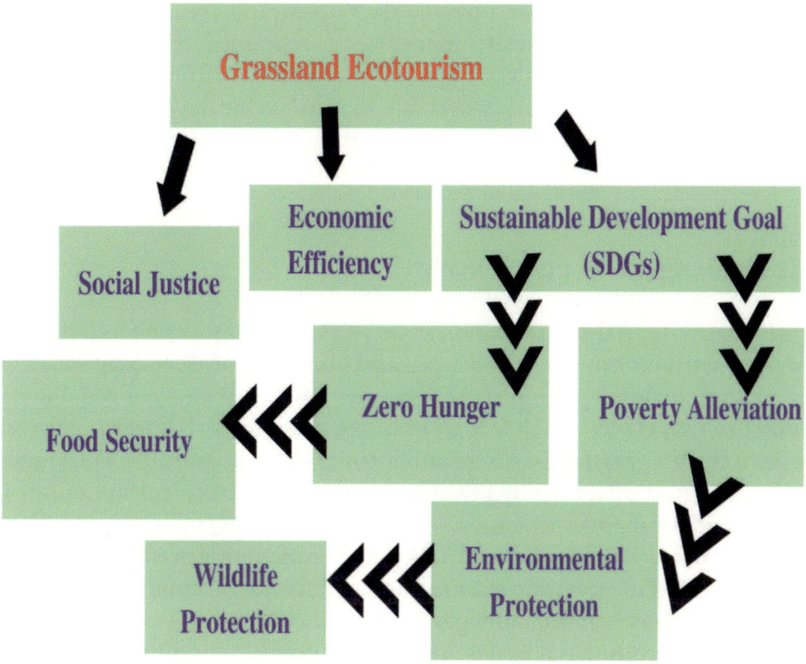

Figure 6.
A variety of advantages that can be achieved by promoting grasslands ecotourism.

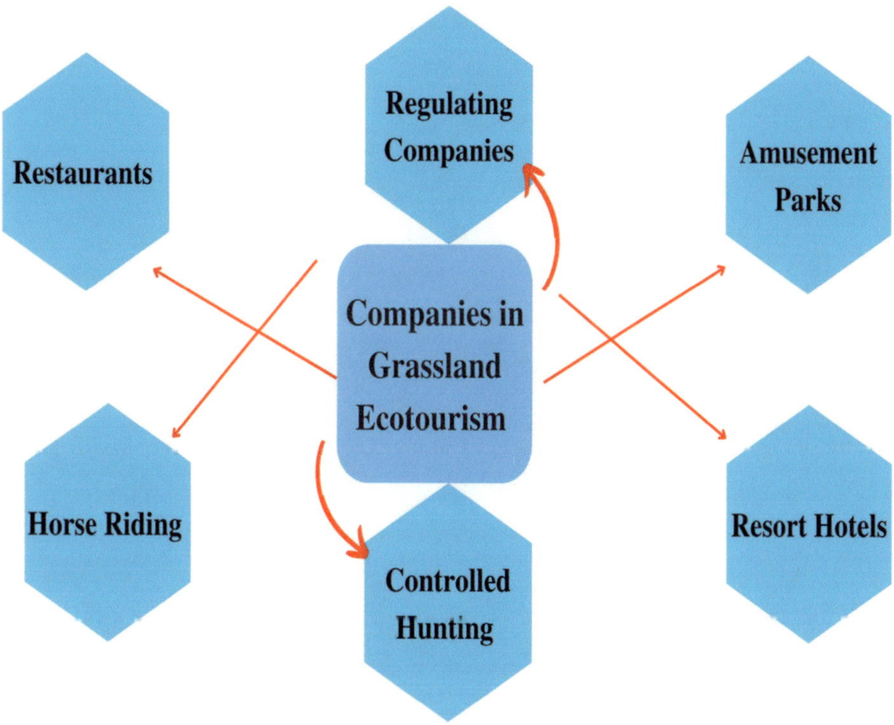

Figure 7.
Different enterprises that may be involved to promote grasslands-based ecotourism and for generating revenues.

development goals of zero hunger, poverty alleviation, food security, and environment protection through community engagement, economic efficiency, social justice, and environment conservation [45]. **Figure 7** summarizes different enterprises that can be involved in producing quality services to tourists for developing grassland ecotourism in the greater benefit of native communities, tourists, grassland resources, and the environment.

7. Future challenges and perspectives

Despite huge potential of grasslands to serve as new centers of ecotourism, this task cannot be without multiple challenges and hurdles. The development of grasslands for ecotourism loses its charms keeping in view the dryness of regions which entails only dry rangelands. Higher temperatures coupled with lesser precipitation as characterized by dry grasslands. These suboptimal climatic conditions seriously compromise the concept of developing grasslands as an alternative income source for local communities and local governments. Besides suboptimal agroclimatic conditions, another pronounced challenge posed to the development of grasslands for ecotourism is the herder's low socioeconomic status. The poor economic conditions of local herders serve as one of the biggest challenges and distort human acts of goodwill. Thus, climate change, environmental issues, and poor economic condition of local herders have significantly counteracted the benefits associated with ecotourism especially related to biomass conservation. As experienced in the dry grasslands of China, rural community explicitly declined to abide by the set of regulation and policies for grasslands development. Likewise, lack of reforms in the grassland's tenure system has caused the division of public grasslands into fragmented, herders-owned plots

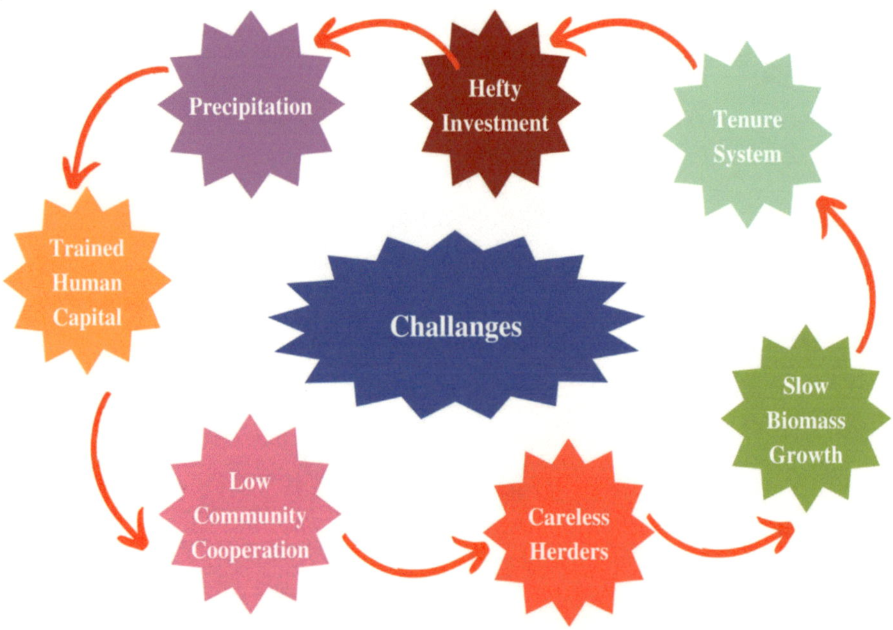

Figure 8.
Different pronounced challenges confronted to the development of grasslands for ecotourism.

which get subject to severe overuse without any management plan. The net result is those herders are never interested and inclined to community cooperation while facing numerous external challenges alone [45–47]. **Figure 8** shows numerous but the most serious challenges confronted to the development of grasslands for ecotourism.

Moreover, other pronounced challenge in the way of developing grasslands for ecotourism is consistent drainage of human capital from rural communities of regions in the vicinity of grasslands. Over time, this situation has led to communities having higher number of aging people having little desire and motivation to strive for developing grasslands. Interestingly, such demographics also result in weak public services and deteriorated infrastructure [48]. Thus, living in socioeconomically declining community, local herders are not able to realize the potential advantages of community-based ecotourism. Hence, without addressing these socioeconomic hurdles, it may be overoptimistic theory to endorse positive social-ecological outcomes of ecotourism.

Future perspective of developing grasslands for ecotourism depends on extensive research and policy initiatives pertaining to ES and food security as grasslands can be utilized for attaining green forage and preserved fodder. This can lead to boost the meat production alongside many other ES. By integrating grasslands into farming systems and land-use decisions locally and regionally, their potential to contribute to functional landscapes and to food security and sustainable livelihoods can be greatly enhanced through research and facilitating different stakeholders. Semi-natural grassland is a product of human management and requires planned livestock grazing or hay cutting in order to appropriately maintain them, and might be encroached by invasive shrubs and trees if not properly cared of which is bound to decrease grasslands value for ecotourism. Last but not least, future research needs to focus on improved grasslands which are the pastures developed from plowing and sowing of non-native grasses with high production potential. Agronomists and horticulturists must redirect their research for formulating optimized practices for ensuring robust growth of native vegetation in grasslands through artificial fertilization and thereafter suggesting biologically viable management package for improved grasslands in order to obtain diversified ES.

8. Conclusions

Grasslands (both natural and developed) are one of the most essential elements of sustainable ecosystem for having multifunctional roles in the perspectives of economics, ecology, sociology, and cultural ethics. Grasslands need to be appropriately managed in order to sustain the grass-based ecosystems and surrounding communities that rely on them for their livelihood. It has been inferred that grasslands development can assist in raising of milk and draught animals paves the way to provide farmers with sufficient income and that too with relatively meager capital investment. In contrast, regions where grasslands are not productive enough to provide sufficient food products on competitive basis owing to adverse environmental conditions, the only way is to create a grassland tourism activity. The development of grasslands for ecotourism offers valuable recreation and generates tourism-associated business activities. The development of grasslands for ecotourism must be done by keeping in view that one of the main purposes of tourists is to spend a holiday and enjoy from a range of site seeing, services, and amenities in a natural environment free from pollution and hustle bustle of cities life. Ecotourism also holds potential to attract typical tourist who are looking for organic products (wine, fruits, olive oil, honey, cheese, meat, milk, etc.). Grasslands must be managed by using sustainable practices that are bound to ensure high diversity of species in multifaceted habitats that attract many species of birds, vertebrates, and invertebrates. The local communities can earn higher profits by selling quality products as consumers perceive that food products coming from organic grasslands are of higher quality. Moreover, the development of grasslands for promoting ecotourism can put a halt to migration of local communities to nearby cities. However, there are numerous challenges in the way of grassland ecotourism such as hefty investment requirement, need of trained human capital, and establishment of a paradigm between grasslands conservation and development activities for tourism. Furthermore, side effects of grasslands ecotourism can only be effectively addressed by systematic evaluation and adopting the appropriate measures. Finally, there are emerging concerns that presence of tourists in large number deteriorate ecosystem functioning of the grasslands. However, a variety of benefits including sociocultural values promotion, exchange of knowledge and ethics, monetary benefits, and conservation of environment advocate development of grasslands for ecotourism.

Author details

Raina Ijaz[1], Nidaa Harun[2] and Muhammad Aamir Iqbal[3]*

1 Faculty of Agriculture, Department of Horticulture, University of Poonch Rawalakot, Pakistan

2 Faculty of Life Sciences, Department of Botany, University of Okara, Pakistan

3 Faculty of Agriculture, Department of Agronomy, University of Poonch Rawalakot, Azad Jammu & Kashmir, Pakistan

*Address all correspondence to: muhammadaamir@upr.edu.pk

References

[1] Iqbal MA. Introductory Chapter: Grasslands Development-Green Ecological Economy and Ecosystem Services Perspectives. London, UK: IntechOpen; 2022. DOI: 10.5772/intechopen.105345

[2] Iqbal MA, Khalid S, Ahmed R, Khan MZ, Rafique N, Ija R, et al. Underutilized Grasses Production: New Evolving Perspectives. London, UK: IntechOpen; 2022. DOI: 10.5772/intechopen.105375

[3] Bovolenta S, Dovier S, Parente G. Dairy production systems in the Italian alpine area. In: Contribution of mountain pastures to agriculture and environment. In: Proceedings of the 16th Meeting of the FAO CIHEAM Mountain Pasture Network. Swiss and ITEP, Poland: ACW; 2011. pp. 143-146

[4] Piasentier E, Martin B. From grass to fork. In: Biala K, Noesberger J, Parente G, Peeters A, editors. Quality production and quality of the environment in the mountain pastures of an enlarged Europe Proceedings of the 13th Meeting of the FAO-CHIEAM, Udine, Italy. Italy: Food and Agriculture Organization; 2006. pp. 109-125

[5] Hopkins A, Holz B. Grassland for agriculture and nature conservation: Production, quality and multifunctionality. Agronomy Research. 2006;**4**(1):3-20

[6] Osterburg B, Isermeyer F, Lassen B, Roder N. Impact of economic and political drivers on grassland use in the EU. Grassland Science in Europe. 2010;**15**:14-28

[7] Erb KHT, Fetzel T, Kastner C, Kroisleitner C, Lauk A, Mayer NM. Livestock grazing, the neglected land use. In: Haberl H, Fischer-Kowalski M, Krausmann F, Winiwarter V, editors. Social Ecology, Human-Environment Interactions. Vol. 5. New York, USA: Springer; 2016. pp. 295-310

[8] Habel JCJ, Dengler M, Janišová P, Török C, Wellstein WM. Europeangrasslandecosystems:Threatened hotspots of biodiversity. Biodiversity and Conservation. 2013;**22**:2131-2138

[9] Bullock JM. Semi-natural grasslands. In: UK NEA, the UK National Ecosystem Assessment. Cambridge, UK: UNEP-WCMC; 2011. pp. 161-196

[10] Lemaire GJ, Hodgson CA. Grassland Productivity and Ecosystem Services. Wallingford, UK: CABI; 2011

[11] Parr CL, Lehmann CER, Bond WJ, Hoffmann WA, Andersen AN. Tropical grassy biomes: Misunderstood, neglected, and under threat. Trends in Ecology and Evolution. 2014;**29**:205-213

[12] Queiroz C, Beilin R, Folke C, Lindborg R. Farmland abandonment: Threat or opportunity for biodiversity conservation? Frontiers in Ecology and the Environment. 2014;**12**:288-296

[13] Suttie JM, Reynolds SG, Batello C. Grasslands of the World. Rome, Italy: FAO; 2005

[14] Pilgrim ES et al. Interactions among agricultural production and other ecosystem services delivered from European temperate grasslands. Advances in Agronomy. 2010;**109**:117-154

[15] Gobster PH, Nassauer JI, Daniel TC, Fry G. The shared landscape: What does aesthetics have to do with ecology? Landscape Ecology. 2007;**22**:959-972. DOI: 10.1007/s10980-007-9110-x

[16] Bonnet X, Shine R, Lourdais O. Taxonomic chauvinism. Trends in Ecology & Evolution. 2002;**17**:1-3. DOI: 10.1016/S0169-5347(01)02381-3

[17] Saunders FP. Seeing and doing conservation differently. Journal of Environment & Development. 2012;**22**:3-24. DOI: 10.1177/1070496512459960

[18] Appleton J. The Symbolism of Habitat: An Interpretation of Landscape in the Arts. Seattle, WA: University of Washington Press; 1991

[19] Chapin FS, Knapp CN. Sense of place: A process for identifying and negotiating potentially contested visions of sustainability. Environmental Science & Policy. 2015;**53**:38-46. DOI: 10.1016/j.envsci.2015.04.012

[20] Archer S. Have southern Texas savannas been converted to woodlands in recent history? American Naturalist. 1989;**134**:545-561

[21] Enquist CAF, Gori DF. Application of an expert system approach for assessing grassland status in the U.S.-Mexico borderlands: Implications for conservations and management. Natural Areas Journal. 2008;**284**:414-428

[22] Loveland TR, Reed BC, Brown JF, Ohlen DO, Zhu Z, Yang L, et al. GLCCD (global land cover characteristics database), version 1.2. Development of a global land cover characteristics data base and IGBP DISCover from 1-km AVHRR data. International Journal of Remote Sensing. 1998;**21**:1303-1330

[23] Brandon K. Ecotourism and Conservation: A Review of Key Issues. Washington, DC, USA: The World Bank; 1996. p. 80

[24] Briske DD, Coppock DL, Illius AW, Fuhlendorf SD. Strategies for global rangeland stewardship: Assessment through the lens of the equilibrium–non-equilibrium debate. Journal of Applied Ecology. 2020;**57**(6):1056-1067. DOI: 10.1111/1365-2664.13610

[25] Yu Q, Peng Y, Liu Z, Li T, Hu J, Huang T, et al. Landscape pattern and tourism aesthetic value of shrubs in the mountainous area of West Sichuan, China. Journal of Landscape Research. 2020;**12**(4):71-82

[26] Ziffer K. Ecotourism: The Uneasy Alliance. Washington, DC, USA: Conservation International; 1989. p. 36

[27] Rahman A. Applıcatıon of GIS in Ecotourısm Development: A Case Study in Sundarbans, Bangladesh [Thesis]. Sundsvall, Sweden: Mid-Sweden University Master of Arts, Human Geography Focusing on Tourism; 2010. p. 79

[28] The International Ecotourism Society. 2023. Available online: https://ecotourism.org/what-is-ecotourism [Accessed: June 6, 2023].

[29] Fennell DA. Ecotourism. fourth ed. United Kingdom: Routledge, Taylor & Francis; 2015. pp. 3-20

[30] Wondirad A. Does ecotourism contribute to sustainable destination development, or is it just a marketing hoax? Analyzing twenty-five years contested journey of ecotourism through a meta-analysis of tourism journal publications. Asia Pacific Journal of Tourism Research. 2019;**24**(11):1047-1065. DOI: 10.1080/10941665.2019.1665557

[31] Gössling S. Ecotourism: A means to safeguard biodiversity and ecosystem

functions? Ecological Economics. 1999;**29**(2):303-320. DOI: 10.1016/S0921-8009(99)00012-9

[32] Wunder S. Ecotourism and economic incentives — An empirical approach. Ecological Economics. 2000;**32**(3):465-479. DOI: 10.1016/S0921-8009(99)00119-6

[33] Stronza AL, Hunt CA, Fitzgerald LA. Ecotourism for conservation? Annual Review of Environment and Resources. 2019;**44**(1):229-253. DOI: 10.1146/annurev-environ101718-033046

[34] Coria J, Calfucura E. Ecotourism and the development of indigenous communities: The good, the bad, and the ugly. Ecological Economics. 2012;**73**(1):47-55. DOI: 10.1016/j.ecolecon.2011.10.024

[35] Das M, Chatterjee B. Ecotourism: A panacea or a predicament? Tourism Management Perspectives. 2015;**14**(1):3-16. DOI: 10.1016/j.tmp.2015.01.002

[36] McKercher B. Academia and the evolution of ecotourism. Tourism Recreation Research. 2010;**35**(1):15-26. DOI: 10.1080/02508281.2010.11081615

[37] Wall G. FORUM: Is ecotourism sustainable? Environmental Management. 1997;**21**(4):483-491. DOI: 10.1007/s002679900044

[38] Li WJ, Li YB. Managing rangeland as a complex system: How government interventions decouple social systems from ecological systems. Ecology and Society. 2012;**17**(1):15. DOI: 10.5751/ES-04531-170109

[39] Ocampo L, Ebisa JA, Ombe J, Geen Escoto M. Sustainable ecotourism indicators with fuzzy Delphi method – A Philippine perspective. Ecological Indicators. 2018;**93**:874-888. DOI: 10.1016/j.ecolind.2018.05.060

[40] Saleh F, Karwacki J. Revisiting the ecotourist: The case of grasslands national park. Journal of Sustainable Tourism. 1996;**4**(2):61-80. DOI: 10.1080/09669589608667259

[41] Li L, Dong Y, Zhang T, Wang H, Li H, Li A. Environmental and social outcomes of ecotourism in the dry rangelands of China. Journal of Ecotourism. 2022, 2022;**22**:1-21

[42] Li A, Wu J, Zhang X, Xue J, Liu Z, Han X, et al. China's new rural "separating three property rights" land reform results in grassland degradation: Evidence from Inner Mongolia. Land Use Policy. 2018;**71**(2):170-182. DOI: 10.1016/j.landusepol.2017.11.052

[43] Maude G, Reading RP. The role of ecotourism in biodiversity and grassland conservation in Botswana. Great Plains Research. 2010;**1**:109-119

[44] Lowder SK, Skoet J, Raney T. The number, size, and distribution of farms, small holder farms, and family farms worldwide. World Development. 2016;**87**:16-29. DOI: 10.1016/j.worlddev.2015.10.041

[45] Sternberg T. Investigating the presumed causal links between drought and dzud in Mongolia. Natural Hazards. 2018;**92**(S1):27-43. DOI: 10.1007/s11069-017-2848-9

[46] Li WJ, Huntsinger L. In pursuit of knowledge: Addressing barriers to effective conservation evaluation. Ecology and Society. 2011;**16**(2):14. DOI: 10.5751/ES-04099-160214

[47] Liao C, Agrawal A, Clark PE, Levin SA, Rubenstein DI. Landscape sustainability science in the drylands:

Mobility, rangelands and livelihoods. Landscape Ecology. 2020;**35**(11):2433-2447. DOI: 10.1007/s10980-020-01068-8

[48] Long H, Li Y, Liu Y, Woods M, Zou J. Accelerated restructuring in rural China fueled by 'increasing vs. decreasing balance' land-use policy for dealing with hollowed villages. Land Use Policy. 2012;**29**(1):11-22. DOI: 10.1016/j.landusepol.2011.04.003

Chapter 4

Mercury Emission from Prescribed Open Grassland Burning in the Aso Region, Japan

Satoshi Irei, Satoshi Kameyama, Hiroto Shimazaki, Asahi Sakuma and Seiichiro Yonemura

Abstract

In every Spring, prescribed grassland burning, so-called Noyaki in Japanese, has been conducted for over a 1000 years by local residents in the Aso region, Japan, for the purpose of grassland conservation because Noyaki prevents invasion of woody plants in the grassland and helps the growth of grasses, which were an important resource of primary industry for roofing materials of houses and livestock feed. Meanwhile, biomass burning is known to be one of the most significant sources of airborne substances including mercury. Taking advantage of the characteristics and resources of the place we live in, we here describe our on-going study for the emission of gaseous mercury from the traditional Noyaki in the Aso region and other grasslands of western Japan. During Noyaki, we sampled and measured gaseous mercury from the Noyaki plumes to better understand mercury emissions and cycles in the local environment. Results showed, on average, 3.8 times higher atmospheric mercury concentrations, demonstrating the emission of gaseous mercury from the Noyaki. The possible origins, novel information the results inferred, and future research direction are discussed in this chapter.

Keywords: Minamata convention on mercury, gaseous mercury, biomass burning, plant uptake, dry deposition, wet deposition

1. Introduction

1.1 Prescribed open grassland burning in the Aso region for the conservation of grassland

The country of Japan consists of 14,125 islands, is located at the longitude between 153°59′12″ and 122°55′57″ and the latitude between 20°25′31″ and 45°33′26″, and has the total area of 377, 974.17 km^2 (**Figure 1a**) [1]. Approximately 126 million people reside in this country [2]. Depending on the region, the climate is divided into sub-tropical, temperate, or boreal and the temperate predominantly covers most of the land. Japan is a green country; forests and grassland account for 66 and 5% of the whole land [3, 4]. The largest grassland in Japan is located in the Aso region, Kumamoto Prefecture.

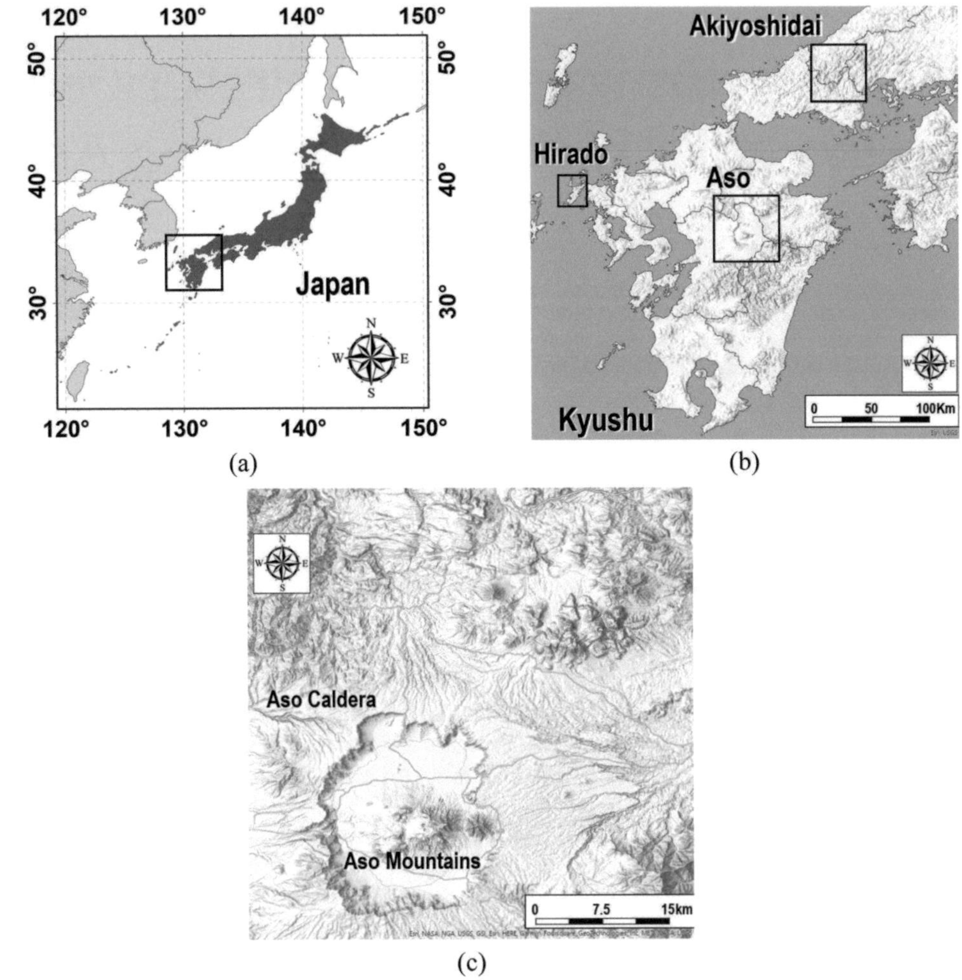

Figure 1.
(a) Map of Japan, (b) Kyushu Mainland, and (c) Aso Region. Map images are from the Elevation World Hillshade (courtesy to the ESRI base map, https://doc.arcgis.com/en/data-appliance/2022/maps/world-hillshade.htm).

Aso is a regional name and it is located nearly in the center of Kyushu Island in western Japan (**Figure 1b**). Geographically, this region consists of a large caldera (463 km^2) with the surrounding area (**Figure 1c**). The altitude is between 400 and 800 m, and the average temperature throughout the year is 13°C approximately [5]. Active volcanoes located in the center of the caldera, "the Aso five mountains", are the landmark of this region. With respect to local governments, the Aso region consists of seven municipalities: Aso city, Oguni town, Minami Oguni town, Takamori town, Ubuyama village, Minami Aso village, Nishihara village. Agriculture, pasture, and tourism are the major industries in these municipals, and the latter two use grasslands. The landscape of the grassland contributes to the local economy. Approximately 1,850,000 tourists in 2018 and 829,000 tourists visited and stayed in the Aso region [6, 7]. This accounts for approximately 23% of tourist visits of Kumamoto Prefecture. Highland breeze in the grassland attracts visitors from all over Japan and overseas.

The grasslands in the Aso region and other local areas such as Hirado in Nagasaki Prefecture and the Akiyoshi-dai plateau in Yamaguchi Prefecture are burnt every Spring (*e.g.*, **Figure 2a** and **b**) after the annual grass dies and dried during the late Fall (**Figure 3**).

Figure 2.
(a and b) Photograph of Noyaki in Aso.

Figure 3.
Photograph of the Aso grassland before Noyaki.

The grassland burning is called "Noyaki" or "Yamayaki" in Japanese (referred to as Noyaki hereafter). According to a scientific study, Noyaki in Aso has been conducted since 10,000 years ago [8]. The purpose of Noyaki in the ancient era was to have open sight for hunting wild animals, but now it has shifted to the prevention of wild forestation of the unused lands, pest control, and reset of annual habitat plants [8, 9]. With this background, there is no doubt that people love Noyaki and it is a symbol of Spring's coming.

1.2 Plant communities in the Aso grassland

The grasslands of the Aso region in Kyushu, Japan, exist both inside and outside of the Aso-Kuju National Park, which covers an area of 72,680 ha [10]. The area of grassland dominated by native species such as *Miscanthus* (*Miscanthus sinensis*) and *Nezasa* (*Pleioblastus chino var. viridis*) within the national park, covering approximately 15,000 ha, approximately 43% of whole grassland in the Aso region [11]. In the grassland, controlled or prescribed burning is conducted as a traditional event every March. This artificial burning event has a significant impact on the regeneration of grassland vegetation and the dominance of certain species. The dominant plant species in the burned grassland include the following [12, 13]: Herbaceous plants of Miscanthus (*Miscanthus sinensis*), bamboo grass (*Pleioblastus chino var. viridis*), Eulalia grass (*Miscanthus transmorrisonensis*), silver grass (*Miscanthus sacchariflorus*), pampas grass (*Miscanthus floridulus*), Japanese yamayuzu (*Fallopia japonica*), Japanese ladybell (*Adenophora triphylla*), bracken fern (*Pteridium aquilinum*); Shrubs of willows (*Salix spp.*), oaks (*Quercus spp.*), sourberries (*Viburnum spp.*), mountain azaleas (*Rhododendron brachycarpum*), painted ferns (*Athyrium niponicum*), hydrangeas (*Hydrangea spp.*), honeysuckles (*Lonicera spp.*), and others. Furthermore, the following plants, which invaded during the time when the mainlands of Japan, Kyushu, and Shikoku islands were connected to the Asian continent, remain as endemic species: "Continental relics": field chickweed (*Stellaria matsudae*), false hellebore (*Veratrum nigrum*), daylily (*Hemerocallis citrina*); "Boreal plants": herbaceous saxifrage (*Saxifraga stolonifera*), golden-rayed lily (*Lilium auratum*), Bistort (*Bistorta officinalis Delarbre subsp. japonica*), lilies (*Lilium spp.*).

1.3 Mercury emission from grassland burning

Crutzen et al. [14] reported the significance of biomass burning as a source of airborne substances in the global atmosphere. Since then extensive emission studies have been done on this subject in fields (*e.g.*, [15–31]), in laboratories (*e.g.*, [32–40]), by satellites (*e.g.*, [40]), and by modeling [41–45]. Published results have also been reviewed (*e.g.*, [46–49]).

Veiga et al. [50] reported indirect observations of gaseous mercury emissions from biomass burning for the first time. Since then gaseous mercury from biomass burning has also been studied in North America [37, 51–53], Africa [54], Europe and Russian Federation [55], South America [56], long-range transport [57–59], and laboratory [38, 40]. To date, it has been reported that biomass burning accounts for ~8% of global mercury emissions to the atmosphere [60]. To the best of our knowledge, the emission of mercury from Noyaki in Japan was reported for the first time by Irei [61, 62]. Its stable mercury isotopic compositions were also reported for the first time worldwide. Even though the emission size may not be as large as wildfires overseas, there is no information available for Noyaki to that date.

Plants are known to uptake mercury from the ground and ambient air (*e.g.*, [63–82]). Sawgrass, a dominant plant species found in the Aso region, is also not an exception. Sawgrass in Florida, relative plant species of Japanese sawgrass, inhales gaseous mercury from the air and fixes it in its own body [83]. Thus, there is no surprise even if Japanese Noyaki emits gaseous mercury to the atmosphere.

The Japanese Ministry of Environment recently updated the domestic emission inventory of mercury for the fiscal year of 2020 (**Figure 4**) [84], showing that ~87% of mercury, out of 10.7 tons of annual mercury emissions, originates from anthropogenic sources. Only volcanic emissions, accounting for ~13%, are the source other than anthropogenic sources. In order to better the mercury cycle in the natural environment, we should not miss any significant source, and Noyaki is one of the veiled domestic emission sources. How much of gaseous mercury was emitted during Noyaki? From where did the emitted mercury come from? Finding answers to these open questions will make our mercury cycle studies progress.

In general, the total emission (TE, g) of chemical species of interest from biomass burning can be estimated using a common mass balance approach.

$$\mathrm{TE} = \mathrm{FL} \times \mathrm{BA} \times \mathrm{CE} \times \mathrm{EF} \quad (1)$$

where a fuel load or FL, also used to be referred to as a "biomass density", is a dried mass of biomass per unit area (typically in kg m^{-2}), burned area or BA is the total area burnt (m^2), combustion efficiency or CE is the fraction of converted biomass carbon to airborne carbonaceous products, such as carbon dioxide (CO_2), and an emission factor or EF is the quantity of substance of interest emitted per kg of biomass burnt (g kg^{-1}) [18, 47, 51, 60, 85]. FL can be estimated according to the information on the biomass

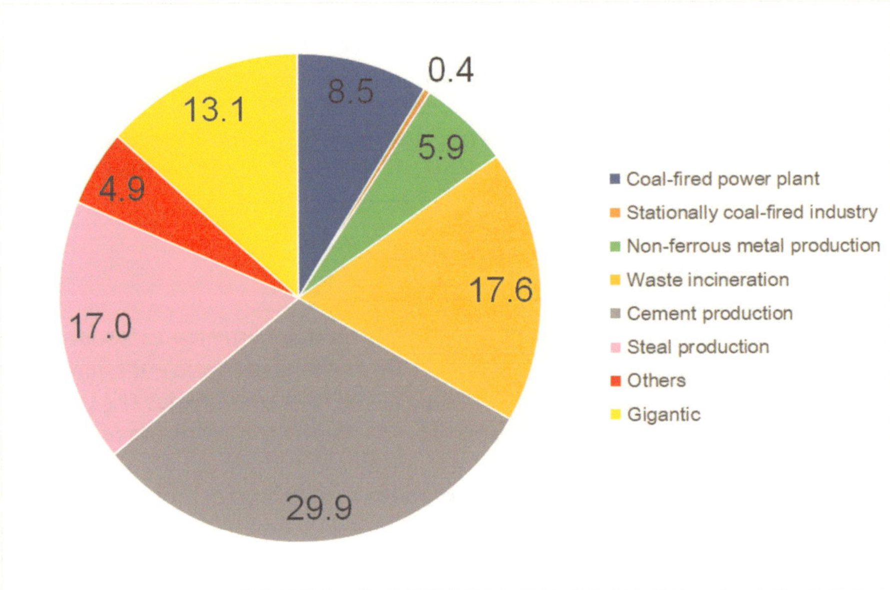

Figure 4.
Domestic Hg emission inventory. The chart was produced according to the information that the Japanese Ministry of Environment has published for the year of 2020 (website). The values are relative to the total domestic Hg emission, 10.7 tons y^{-1}.

density map [41], but it can also be measured directly (*e.g.*, [24–26, 60]). For a large area, FL is highly uncertain [42]. BA is primarily achieved using remote sensing technology through two methods [86]: The approach involving calculation of the extent of the burned area through the comparison of two images taken before and after the fire [87, 88] and the approach involving time-series analysis of regularly captured satellite images from the real-time monitoring of the fire [89] using the MODIS satellite (https://modis.gsfc.nasa.gov/data/dataprod/mod14.php). A CE is hard to know in a real fire, but a modified CE (MCE), the excess concentration (denoted as Δ hereafter, which is the concentration during Noyaki subtracted by the background concentration) ratio of CO_2 to the sum of excess concentrations of carbon monoxide (CO) and CO_2 ($\Delta CO + \Delta CO_2$), has been used as an alternate index of CE [18, 90]. An EF for a substance of interest is a key parameter depending on the types and moisture content of the fuel and burning conditions (smoldering and flaming fires). Of the parameters and variable in Eq. (1), EFs have been most intensively discussed in biomass-burning emission studies (*e.g.*, [46]). An EF can be further broken down to as follows [18, 46]:

$$EF = ER \times \frac{MW_i}{12} \times CC \times 1000 \tag{2}$$

ER is an excess mole concentration ratio of the targeted substance to the sum of the excess carbon mole concentrations of carbonaceous (reference) substances, which are measured in field studies, MW_i is the molecular weight of substance of interest (200.59 g mol^{-1} in the case of mercury), 12 is the atomic mass of carbon, and CC is the carbon content of fuel, which is often assumed as 0.5 [18, 36], and 1000 is a conversion factor for the product in g base to kg base. It should be noted that the sum of the excess concentrations of carbonaceous species is often approximated as the sum of ΔCO and ΔCO_2 or the excess mole concentration of CO_2 divided by MCE.

1.4 Sampling and measurement

In the Aso region, there are a number of stock farms and they possess own grasslands. In every Spring they burn the grasslands to restore the perennial plants. Thus, occurrences of prescribed fires are spotted over the region. Approximately 160 km^2 of the Aso grassland is burnt in total every Spring [91].

Field sampling and measurements for chemical substances emitted from Noyaki can be conducted by either airborne or on the ground. The former method (*i.e.*, use of aircraft) is often chosen in biomass burning studies overseas because open biomass burnings, mostly wildfires, are large and data gained during flights over plumes can represent the emission [46]. It is also because airborne sampling and measurements are more safely conducted. However, the high cost, limited accessibility, fast moving speed, etc. of manned aircraft sampling and measurements are overpowered in the Japanese Noyaki studies where only one or few km^2 of grassland is burnt at each stock farm. Meanwhile, ground-based sampling and measurements can be done at fixed measurement stations and/or on ground vehicles. The advantages of ground-based measurements are that instruments and measuring devices can be set more easily and quickly, compared to aircraft measurements. Relatively low cost of use of cars is another advantage. However, the success of ground-based measurements strongly relies on the weather conditions, such as wind speed and directions, proximity to the fires and plumes, landscapes of grasslands, paved roads one can drive vehicles on, etc. Depending on how quickly fire spreads, safety is another factor that one must consider. In our field studies, we have chosen ground-based sampling and measurements using vehicles.

In biomass burning studies, measurements of CO and CO_2 are the base because the major products of biomass burning are CO_2 and CO, accounting for more than 95% of products in general, and their ratio often indicates burning conditions, such as smoldering and flaming. CO_2 and CO are also contained in background air, therefore, their excess concentrations from their background level need to be checked for the evaluation of burning conditions.

Gaseous mercury concentrations can be measured either by automated mercury analyzers or the combination of conventional amalgam trap sampling and its offline analysis using a cold-vapor atomic fluorescent or absorbance spectrometer.

For stable mercury isotope analysis, nano gram order of total gaseous mercury (TGM) needs to be collected. This collection can be done through either chemical traps [92], multiple commercially available gold amalgam sampling tubes [80, 93], or large volume gold amalgam sampling tubes, namely BAuTs [62]. Collected TGM was then converted to gaseous elemental mercury (GEM) by heating, and the converted GEM is oxidized to Hg^{2+} and captured in sulfuric acid/permanganate mixture or reversed aqua resia mixture. Prepared sample solutions were then analyzed by a multi-collector inductively coupled plasma mass spectrometer or MC-ICP-MS.

1.5 Potential of Noyaki studies in western Japan

Since the start of our Noyaki emission studies in 2019, solid evidence of mercury emission has been confirmed [61, 62]. The papers also report the similarly fractionated $\delta^{x}Hg$ values of the excess TGM to the fractionated $\delta^{x}Hg$ values of mercury found in plant species reported by others, implying that the excess TGM was likely supplied from the grassland plant. The pilot studies above gave us an overview of what was happening during the Noyaki events and raised some intriguing open questions: How much was TGM produced and emitted to the atmosphere from Noyaki in this region? What was the origin of TGM emitted from Noyaki?

The first question is the primary one that anyone will have. This estimation has not ever been done, thus, the source has not been included yet in the domestic mercury emission inventories of gaseous mercury that the Ministry of Environment Japan has ever reported. Under the current circumstance that a number of nations implement the regulation on man-made mercury use, namely the Minamata Convention on Mercury, this evaluation is worthwhile and contributes to update the mercury budget. It is also interesting how variable "the annual routine emissions" are. Such information is new and can be gained only from the prescribed burning conducted routinely. The information can be extended to the fate of atmospheric mercury, which can be applicable not only to the local environment but global one as well.

The second question delves into the mercury cycle in detail. Identifying the origin (s) will help us to better understand phenomena occurring on mercury in the natural environment, which contributes to the intellectual and novel input into our current understanding of mercury cycle.

2. Methodology

2.1 Estimation of burnt area

In the following manner, we conducted the process of satellite image analysis. First, we downloaded Landsat-8 surface reflectance and surface temperature products (path = 112, row = 037) observed on April 9, 2023, from the United States Geological Survey Earth Explorer website (https://earthexplorer.usgs.gov/). For the purpose of creation of a spatial distribution map of prescribed burns, we adopted a supervised classification approach. In this classification method, we curated ground reference data for two classification categories: "burned" and "unburned." In the next step, we selected three vegetation groups: "pastureland," "*Miscanthus sinensis* (including *Pleioblastus chino var. viridis*) stand", and "*Miscanthus sinensis* (excluding *Pleioblastus chino var. viridis*) stand" using a 1:25,000 scale existing vegetation map. To reduce spatial bias in the ground reference data, we determined 130 reference points from each group. For the three selected vegetation groups (130 ground reference data points each × 3), we randomly allocated "training data (100)" and "validation data (30)" for each vegetation group. We adopted the Random Forest algorithm, a type of machine learning algorithm with a proven track record in land use and land cover classification using satellite imagery. This algorithm was executed using the random Forest package (version 4.7.1.1) in R (version 4.2.1). For the classification results of the controlled burns' spatial distribution map, we evaluated the accuracy with overall accuracy and a Kappa coefficient.

2.2 Fuel loads and emission factor

For wildfire studies, fuel load information can be gained from estimation based on CO and CO_2 measurements under the assumption that the fuel burned contains 50% of carbon content in weight (*e.g.*, [18, 36]) or from the actual analysis of habitat plants in the field [73]. The former is a top-down method and practical for cases where sampling biomass representing the habitat plants in the burned area is difficult (*e.g.*, forests with a large diversity of plant species). In contrast, the latter, a bottom-up method of actual

(a)

(b)

Figure 5.
Photograph of example of sampling aboveground grass in a 1 × 1 m quadrat: (a) before the grass was reaped, and (b) after the grass was reaped.

measurements of biomass, is ideal. For the Aso grassland, this ideal case can be applied because the composition of plant species is simple; only two plant species, *Miscanthus sinensis* and *Pleioblastus chino var. viridis*, predominantly occupy the grassland burned. This makes the retrieval of aboveground fuel loads and chemical contents of the biomass pool feasible.

Fuel load, the mass of biomass per unit area, was obtained experimentally in the Aso grassland study. In the grassland, a 1 × 1 m square quadrat was defined using plastic rods, then the plants and litters within the quadrat were sampled (**Figure 5**). The plant samples were then brought to the laboratory, and dried in the air-conditioned room (24°C and 24% for temperature and humidity, respectively) until the masses were stabilized. Their masses and chemical contents were then measured. If combustion completeness, CE = 1, is given, an EF for a substance of interest will be equivalent to the content in the biomass, thus, the EF will be easily gained from this bottom-up approach.

2.3 Sampling and measurement of airborne chemical species

In our emission study in Aso, sampling and measurements for airborne chemical species were conducted in a vehicle. In-situ measurements for CO and CO_2 were made by a CO analyzer (Model 48C, Thermo Fisher Scientific.) and a CO_2 analyzer (LI-810, LI-COR Corp., Lincoln, NB, U.S.A.). For TGM measurements, plume gas was drawn through a gold-coated sand trap (N-160, Nippon Instruments Corp., Osaka, Japan) at the rate of 0.5 L min^{-1} using a mini-air pump (MP-W5P, Shibata Scientific Technology Ltd., Souka, Japan), and the tube samples were brought to the laboratory and analyzed for TGM by cold-vapor atomic fluorescent spectroscopy (CV-AFS, Nippon Instruments Corp.). TGM for stable mercury isotope analysis was captured through a BAuT sampling tube [61, 62] at the flow rate of 80 L min^{-1}. After the sampling, the BAuT tube was brought to the laboratory, then the TGM captured was converted to oxidized mercury (II) in 40% reversed aqua resia solution. The prepared solution samples were then analyzed by an MC-ICP-MS (Neptune Plus, Thermo Fisher Scientific GmbH, Bremen, Germany) for the isotopic composition.

$$\delta^{x}Hg(‰) = \left[\frac{\left(\frac{^{x}Hg}{^{198}Hg}\right)_{sample}}{\left(\frac{^{x}Hg}{^{198}Hg}\right)_{3133}} - 1\right] \times 1000 \qquad (3)$$

where x stands for the stable mercury isotope with mass x, and the bracketed isotope ratios with subscripts "sample" and "3133" indicate the stable mercury isotope ratios of mass x relative to the mass 198 for the sample and SRM 3133 (NIST), respectively.

Either a Teflon-coated glass fiber filter (Pallflex, Emfab, Pall Corp., Port Washington, NY, U.S.A.) or quartz fiber filter (Pallflex, Tissuequartz, Pall Corp.) installed in filter holder (Innovation NILU AS, Kjeller, Norway) or capsule filter (Balston, Parker Hannifin Corp., Cleaveland, IL, U.S.A.) was attached to all the inlets for sampling and measurements described above.

Instruments and devices referred above were loaded to a vehicle, together with batteries for their power supplies, then plumes from Noyaki were chased. When the car was drove into plumes the car was stopped in the plumes and the engine was stopped until the plume is gone. The sampling and measurements were continued until the Noyaki of the day there ended.

3. Results and discussion

3.1 Burned area

There have been previous efforts that estimate the extent of BA in the Aso region using satellite imagery [94]. However, the estimates were based on the conventional maximum likelihood method and the reliability of the estimates was not evaluated. In our study, we employed the Random Forest algorithm, which is a machine learning method with better classification performance than the maximum likelihood method, and then evaluated the reliability of classification results in terms of overall accuracy and Kappa coefficient. The image analysis results (spatial distribution map of prescribed burns) were very accurate, with an overall accuracy of 0.972 and a Kappa coefficient of 0.944. This analysis demonstrated the potential and practical utility of high-precision spatial distribution estimation for prescribed burn monitoring using satellite imagery. Within the boundaries of individual pasturelands, areas that have undergone prescribed burns and those that have not were intermixed. The obtained data were used for more precise determination of the burnt area on more fine scale, and currently, boundaries are under the cross check. For future work, estimating burned biomass in prescribed burn monitoring, the capability of satellite image analysis covering extensive areas (enabling spatial distribution understanding) will be highly effective and essential (**Figure 6**).

3.2 Measurements

3.2.1 Aboveground fuel loads

Aboveground fuel loads we found in the Aso region are listed in **Table 1**. The table also shows fractions of two major habitat plant species, *Miscanthus sinensis* and *Pleioblastus chino var. viridis*. Results showed that the fuel loads were highly variable, from 0.6 to 1.58 kg m^{-2}. The compositions at most of the locations, except Minamioguni 1, were split by the two plant species. Minamioguni 1 was surrounded by trees, and, therefore, the most contributing, biofuel there was litter, which was included as other. Compared to the fuel loads reported to date, such as for African savanna (0.36–0.48 kg m^{-2} [25], 0.20–2.5 kg m^{-2} [95]), Austrarian savanna (0.56 kg m^{-2} [24], 0.11–0.74 kg m^{-2} [95]), North American meadow (0.4 kg m^{-2}, [52]), and central and south American savanna (0.30–1.38 kg m^{-2} [95]), our observations were in compatible level.

3.2.2 TGM concentration

Observed atmospheric concentrations of TGM in the background air and Noyaki plumes were on average (± SD) 1.4 ± 0.5 and 5.2 ± 3.6 ng m^{-3}, respectively. TGM concentrations varied substantially (**Figure 7**) and the variation was attributed to the proximity to the fire. With consideration of this atmospheric dilution, we are now confident that Noyaki indeed emits TGM into the atmosphere. As stated earlier, this emission has not been included in the Japanese domestic emission inventory that the Ministry of Environment Japan published, therefore, the emission source needs to be evaluated for a better understanding of the mercury cycle in the natural environment.

3.2.3 CO and CO_2 measurements

We defined the background concentrations of CO and CO_2 as 0 ppmv and 422 ppmv, respectively. Therefore, for our data, the subtraction of these concentrations

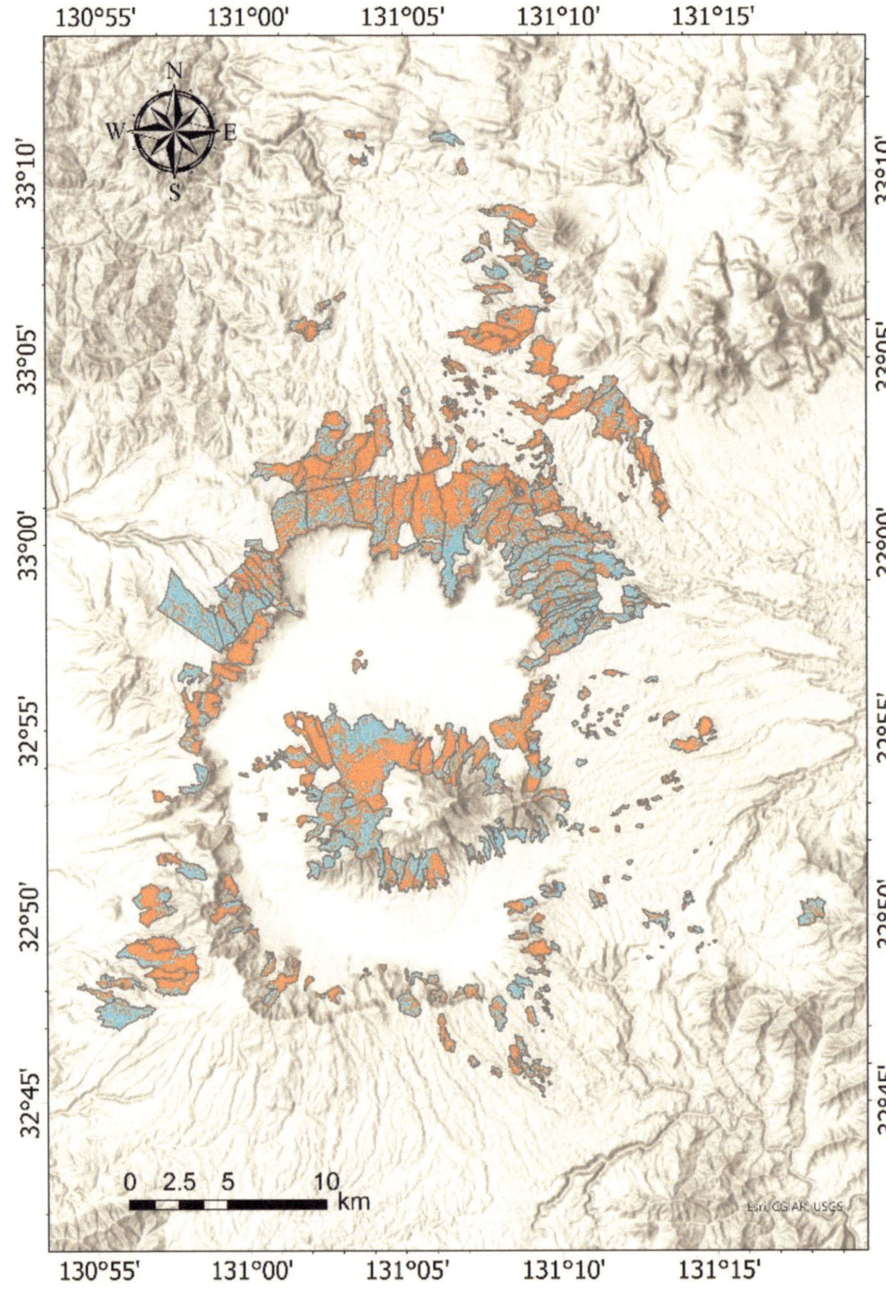

Figure 6.
Overlaid images of a landscape map and burned (red) and unburned (light green) grasslands of the Aso region identified from the satellite images. Sources of the landscape map: World Hillshade, Esri, Airbus DS, USGS, NGA, NASA, CGIAR, N Robinson, NCEAS, NLS, OS, NMA, Geodatastyrelsen, Rijkswaterstaat, GSA, Geoland, FEMA, Intermap, and the GIS user community.

was applied due to their mixing ratios observed in the background air. Using the results obtained from the field study at the Akiyoshidai National Park, we show the overview of this research project.

Sampling location	Fuel load (kg m^{-2})	*Miscanthus sinensis* (%)	*Pleioblastus chino var. viridis* (%)	Other (%)
Hirado	1.09	n.a.	n.a.	n.a.
Nishihara	1.58	n.a.	n.a.	n.a.
Minamioguni 1	0.60	29.5	31.5	39.0
Minamioguni 2	0.74	67.6	22.9	9.5
Kamitajiri	0.96	27.4	71.6	0.9
Takenohata 1	0.96	88.2	11.8	0.0
Takenohata 2	0.73	64.3	30.9	4.9

n.a. = data are not available.

Table 1.
Fuel loads in the grassland of Aso and fractions of the two major habitat plant species, Miscanthus sinensis *and* Pleioblastus chino var. viridis.

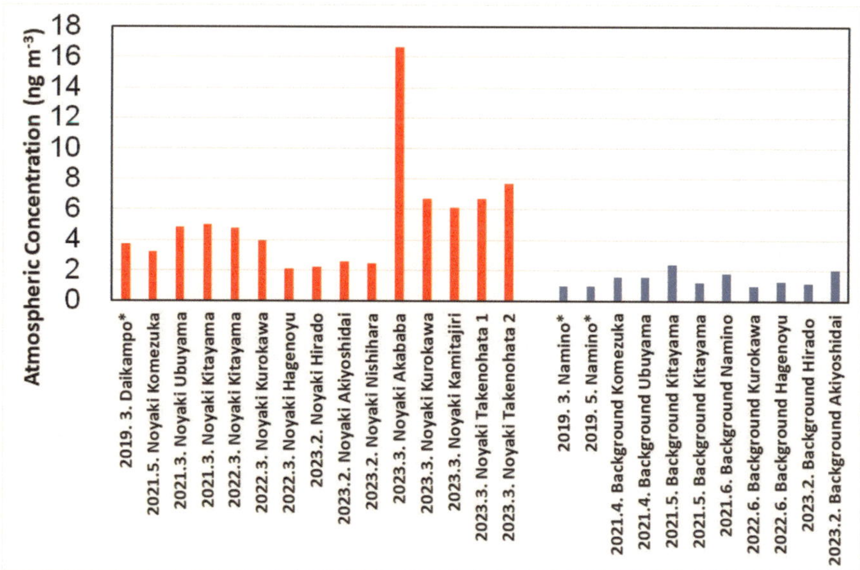

Figure 7.
Average atmospheric concentrations of TGM observed during the Noyaki (red bar) and ordinary days (blue bar) in the period of 2019–2023.

The ground-based measurements of CO and CO_2 showed the complexity of the combustion state, flaming and smoldering. For example, observations for CO and CO_2 at Akiyoshidai in the time period between 10 AM and 12 PM demonstrated their unsynchronized variations in detail, but roughly speaking, their trends were similar (**Figure 8a**). MCEs in this time period varied from 86 to 104% and on average ± SD, 97.2 ± 1.3% (**Figure 8b**). The fractions of CO relative to the sum of CO and CO_2 were on average ± SD, 2.9 ± 4.0%. The scatter plot of ΔCO against ΔCO_2 showed a high correlation (r^2 = 0.768) with a slope of 0.02, indicating a proportional relationship and nearly complete combustion (**Figure 9**).

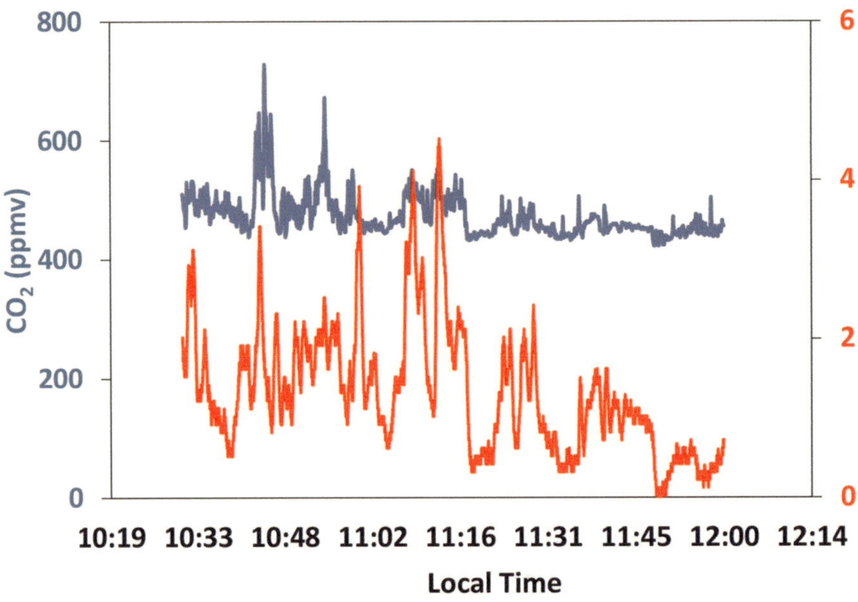

(a)

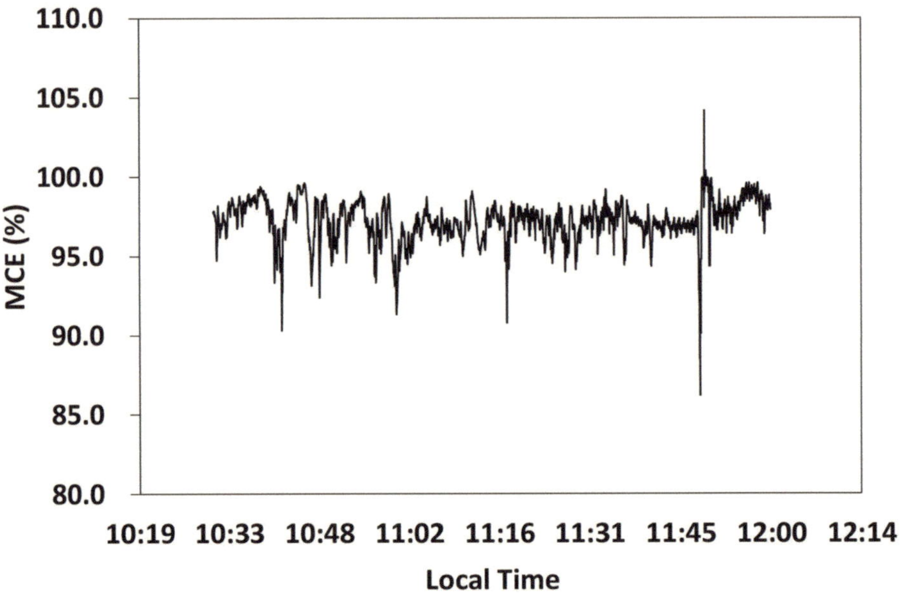

(b)

Figure 8.
Time series plot of (a) volume-based CO_2 (blue) and CO (red) mixing ratios, and (b) modified combustion efficiency (i.e., $\Delta CO_2/(\Delta CO + \Delta CO_2)$) *observed during the prescribed grassland burning at the Akiyoshidai National Park.*

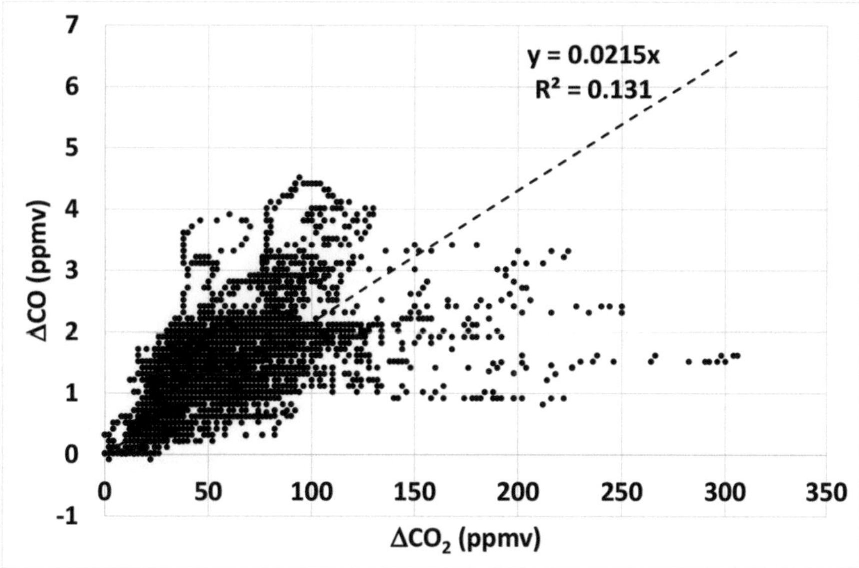

Figure 9.
Scatter plot of excess CO and CO_2 mixing ratios (ΔCO and ΔCO_2) observed during the prescribed grassland burning at the Akiyoshidai National Park.

3.2.4 Excess concentration ratio, emission factor, and Total mercury emission

ERs for $\Delta TGM/(\Delta CO_2 + \Delta CO)$ and $\Delta TGM/\Delta CO$ are approximately 1.6×10^{-9} and 0.56×10^{-7}, respectively. According to Eq. (2) with the assumptions that the carbon content of fuel was 50%, the majority of this carbon was converted to CO_2 and CO, and CO_2:CO was 0.97:0.03 (discussed in subsection 3.2.3 for MCE), the corresponding EFs estimated from the ERs for $\Delta TGM/(\Delta CO_2 + \Delta CO)$ and $\Delta TGM/\Delta CO$ shown above are 13.2×10^{-6} and 14.0×10^{-6} g Hg kg^{-1}, respectively. The values are significantly low, compared to other EFs reported for the grassland fires ranging from 38 to 510×10^{-6} g Hg kg^{-1} ([51], references therein). The results were obtained during the emission study in Akiyoshidai, therefore, we applied these EFs to the prescribed fire there. Given the FL of 1.0 kg m^{-2} with the 11.4 km^2 of burned area reported [96]. The estimated Hg emission from there is 150.5 and 159.6 g y^{-1}, which is approximately 0.02% of the annual Japanese domestic emission of TGM that the Ministry of Environment Japan reported.

3.2.5 Stable mercury isotope ratios

A limited number of TGM samples collected using BAuTs were analyzed for stable mercury isotope ratios. In order to characterize the isotopic compositions we evaluated if the observed isotopic compositions were reflected by mass dependent fractionation (MDF) or mass independent fractionation (MIF) using the Eq. (4) [97].

$$\Delta^x Hg(‰) \approx \delta^x Hg - \left(\delta^{202}Hg \times \beta_x\right) \quad (4)$$

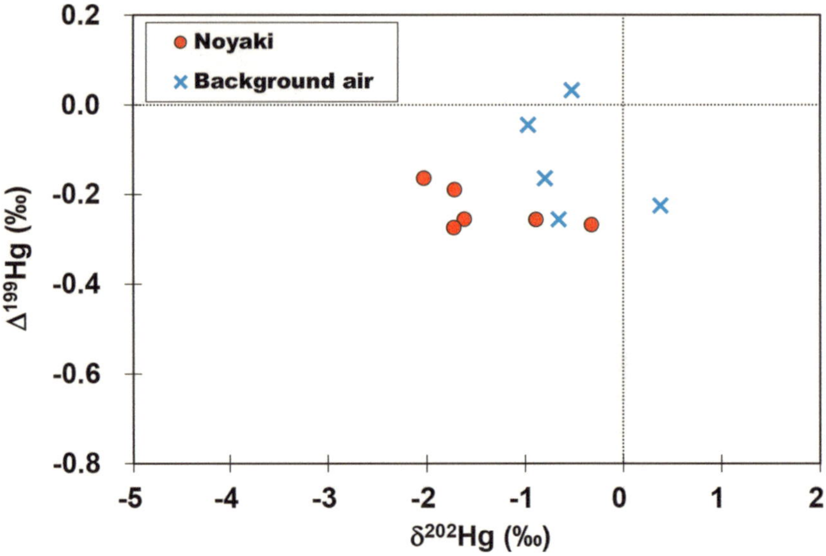

Figure 10.
Stable isotope ratios of TGM collected from the Noyaki plume and background air. The figure was reproduced from the previously published data [60]. See the text and Eqs. (3) and (4) for the definitions of these isotope ratios.

where β_x is the mass-dependent fractionation factor for mass x (0.2520, 0.5024, 0.7520, and 1.4930 for mass 199, 200, 201, and 204, respectively).

Results showed that Δ^{199}Hg values for TGM in the Noyaki plumes were on average ($\pm$ SD), $-$ 0.23 $\pm$ 0.05‰, while those for TGM in the background air were on average, $-$ 0.13 $\pm$ 0.12‰, indicating TGM emitted from the Noyaki underwent small MIF and TGM in the background air did not. Furthermore, δ^{202}Hg of TGM from the background air and Noyaki plumes showed discrete compositions (**Figure 10**). It is interesting that our observations shown in **Figure 10** resembled the isotopic compositions found in plants, such as rice straw [75], foliage [79–81], and in soil [82]. Was atmospheric GEM aspirated from the air together with CO_2, then oxidized and fixed in the plant body? Or was the deposited mercury on the ground taken up through the roots of plants? Answers to those open questions will be found by continuing overall grassland studies, including prescribed biomass burning and wet and dry deposition of mercury as well as mercury content in plant sections and soil.

4. Conclusion

Our study concretely confirmed that the prescribed biomass burning of the grasslands in Aso and other grasslands in western Japan emitted TGM into the atmosphere. The estimation of the emitted TGM during the Noyaki in the Aso region is currently under evaluation. Preliminary estimation of the TGM emission from Noyaki in the Akiyoshidai National Park resulted in 150–160 g TGM emission, which accounts for 0.02% of the whole domestic emission of mercury that the Japanese Ministry of Environment reported. Due to more than 10 times larger area of the Aso grassland than the Akiyoshi-dai grassland, it is possible that the emission of mercury from Noyaki in Aso can rise to 1% or similar, and for whole Noyaki in Japan may rise to a few %. The investigation is currently on-going.

δ^{x}Hg values of TGM showed similar values to the reported δ^{x}Hg of the plant uptake TGM, implying the atmospheric mercury can be the origin. However, our current dataset could not identify whether or not the plant uptake of TGM from the air or wet and/or dry depositions contributed to the grassland mercury. Further detailed study combined with atmospheric monitoring in the grassland may help us to find the answer.

Acknowledgements

The authors thank The Sumitomo Foundation (Grant number 2230266) and the National Institute for Minamata Disease (NIMD) for the financial support of the research on mercury emissions from the grassland burning. The authors also thank the assistant Nanami Yamamoto from the NIMD for assisting our laboratory work in this project.

Conflict of interest

The authors declare no conflict of interest.

Author details

Satoshi Irei[1]*, Satoshi Kameyama[2], Hiroto Shimazaki[3], Asahi Sakuma[3] and Seiichiro Yonemura[4]

1 Department of Environment and Public Health, National Institute for Minamata Disease, Kumamoto, Japan

2 Biodiversity Division, National Institute for Environmental Studies, Ibaraki, Japan

3 Department of Civil Engineering, National Institute of Technology, Kisarazu College, Chiba, Japan

4 Faculty of Bioresource Sciences, Prefectural University of Hiroshima, Hiroshima, Japan

*Address all correspondence to: satoshireinimd@gmail.com

References

[1] Geospatial Information Authority of Japan. Provision of Maps Clearly Showing Territory of Japan [Internet]. Available from: https://www.cas.go.jp/jp/ryodo_eg/torikumi/giaj.html [Accessed: September 7, 2023]

[2] Statistics Bureau of Japan. Current Population Estimates as of October 1, 2022 [Internet]. Available from: https://www.stat.go.jp/english/data/jinsui/2022np/index.html [Accessed: September 7, 2023]

[3] Ministry of Environment, Japan. White Paper; Annual Report on the Environment, the Sound Material-Cycle Society and Biodiversity in Japan. 2022

[4] National Agriculture and Food Research Organization. Available from: https://www.naro.go.jp/project/results/laboratory/nilgs/2012/420b0_05_08.html

[5] Ministry of Environment Kyushu Regional Environment Office, 2003, The Handbook of the Grassland of Aso (in Japanese). Revised March, 2007

[6] Kumamoto Prefecture. Kumamoto Prefecture Tourism Statistics 2018 (in Japanese) [Internet]. 2020. Available from: https://www.pref.kumamoto.jp/uploaded/life/6411_126374_misc.pdf [Accessed: September 7, 2023]

[7] Kumamoto Prefecture. Kumamoto Prefecture Tourism Statistics 2021 (in Japanese) [Internet]. 2022. Available from: https://kumamoto.guide/files/af7bd824-5c2d-4038-bd89-4284c2525321.pdf [Accessed: September 7, 2023]

[8] Sasaki A, Sasaki N. Chapter 8: The history of grassland and human activity in the Aso Kuju region from the perspective of phytolith, pollen, and fine particulate carbon (in Japanese). In: Sato H, Kenji I, editors. The Environmental History of Wildland. Vol. 2, the Series 35000 Years of – Environmental History of Human and Nature. Tokyo: Bunichisogou shuppan; 2011. pp. 169-182. ISBN 978-4-8299-1196-9

[9] Toda Y, Vegetation and Landscape of Aso. Special feature: The natural environment and landscape of Kyushu. Journal of Landscape Architecture. 1994; **57**(4):338-345. (In Japanese)

[10] Aso-Kujyu National Park. Available from: https://www.env.go.jp/park/aso/

[11] Aso Grassland Regeneration Executive Office. Chapter 2: Characteristics of Aso Grassland (in Japanese). In: The 3rd Aso Grassland Regeneration Plan [Internet]. 2021. Available from: https://www.aso-sougen.com/learn/files/mgz_03.pdf [Accessed: 2023]

[12] Aso Pedia. Plants of Aso (in Japanese) [Internet]. 2015. Available from: http://www.aso-dm.net/?%E9%98%BF%E8%98%87%E3%81%AE%E6%A4%8D%E7%89%A9 [Accessed: July 14, 2023]

[13] Aso Geopark Promotion Council. Flora and Fauna of Aso (in Japanese). In: Aso Geopark Guidebook [Internet]. 2012. pp. 24-30. Available from: http://www.aso-geopark.jp/mainsites/mainsite08.html [Accessed: July 14, 2023]

[14] Crutzen PJ, Heidt LE, Kransnec JP, Pollock WH, Seiler W. Biomass burning as a source of atmospheric trace gases CO, H_2, N_2O, NO, CH_3Cl, and COS. Nature. 1979;**282**(15):253-256

[15] Hegg DA, Radke LF, Hobbs PV. Ammonia emissions from biomass burning. Geophysical Research Letters. 1988;**15**(4):335-337

[16] Cofer WR III, Levine JS, Winstead EL, Stocks BJ. Gaseous emissions from Canadian boreal forest fires. Atmospheric Environment. 1990;**24A**(7):1653-1659

[17] Cofer WR III, Levine JS, Winstead EL, Lebel PJ, Koller AM, Hinkle CR. Trace gas emissions from burning Florida wetlands. Journal of Geophysical Research. 1990;**95**(D2):1865-1870

[18] Yokelson RJ, Goode JG, Ward DE, Susott RA, Babbitt RE, Wade DD, et al. Emissions of formaldehyde, acetic acid, methanol, and other trace gases from biomass fires in North Carolina measured by airborne Fourier transform infrared spectroscopy. Journal of Geophysical Research. 1999;**104**(D23): 30109-30125

[19] Yokelson RJ, Crounse JD, DeCarlo PF, Karl T, Urbanski S, Atlas E, et al. Emissions from biomass burning in the Yucatan. Atmospheric Chemistry and Physics. 2009;**9**:5785-5812

[20] Burling IR, Yokelson RJ, Akagi SK, Urbanski SP, Wold CE, Griffith DWT, et al. Airborne and ground-based measurements of the trace gases and particles emitted by prescribed fires in the United States. Atmospheric Chemistry and Physics. 2011;**11**: 12197-12216

[21] Akagi SK, Burling IR, Mendoza A, Johnson TJ, Cameron M, Griffith DWT, et al. Field measurements of trace gases emitted by prescribed fires in southern US pine forests using an open-path FTIR system. Atmospheric Chemistry and Physics. 2014;**14**:199-215

[22] Andreae MO, Artaxo P, Beck V, Bela M, Freitas S, Gerbig C, et al. Carbon monoxide and related trace gases and aerosols over the Amazon Basin during the wet and dry seasons. Atmospheric Chemistry and Physics. 2012;**12**: 6041-6065

[23] Hurst DF, Griffith DW, Carras JN, Williams DJ, Fraser PJ. Measurements of trace gases emitted by Australian savanna fires during the 1990 dry season. Journal of Atmospheric Chemistry. 1994; **18**:33-56

[24] Hurst DF, Griffith DWT, Cook GD. Trace gas emissions from biomass burning in tropical Australian savannas. Journal of Geophysical Research. 1994; **99**(D8):16441-16456

[25] Shea RW, Shea BW, Kauffman JB, Ward DE, Haskins CI, Scholes MC. Fuel biomass and combustion factors associated with fires in savanna ecosystems of South Africa and Zambia. Journal of Geophysical Research. 1996; **101**(D19):23551-23568

[26] Ward DE, Hao WM, Susott RA, Babbitt RE, Shea RW, Kauffman JB, et al. Effect of fuel composition on combustion efficiency and emission factors for African savanna ecosystems. Journal of Geophysical Research. 1996; **101**(D19):23569-23576

[27] Cofer WR, Levine JS, Winstead EL, Cahoon DR, Sebacher DI, Pinto JP, et al. Source compositions of trace gases released during African savanna fires. Journal of Geophysical Research. 1996; **101**(D19):23597-23602

[28] Le Canut P, Andreae MO, Harris GW, Wienhold FG, Zenker T. Airborne studies of emissions from savanna fires in southern Africa 1. Aerosol Emissions Measured with a Laser Optical Particle Counter. 1996;**101**(D19):23615-23630

[29] Andreae MO, Andreae TW, Annegarn H, Beer J, Cachier H, Le Canut P, et al. Airborne studies of aerosol emissions from savanna fires in southern Africa: 2. Aerosol chemical composition. Journal of Geophysical Research. 1998; **103**(D24):32119-32128

[30] Echalar F, Gaudichet A, Cachier H, Artaxo P. Aerosol emissions by tropical forest and savanna biomass burning: Characteristic trace elements and fluxes. Geophysical Research Letters. 1995;**22**(22):3039-3042

[31] Yamazaki D, Kajiwara H, Kirii M, Ohira S, Toda K. Direct determination of polycyclic aromatic hydrocarbons in PM2.5 by thermal desorption-GC/MS and analysis of their diurnal/seasonal variations and field burning in Kumamoto (in Japanese). Bunseki Kagaku. 2015;**64**(8):571-579

[32] Yokelson RJ, Susott R, Ward DE, Reardon J, Griffith DW. Emissions from smoldering combustion of biomass measured by open-path Fourier transform infrared spectroscopy. Journal of Geophysical Research [Atmosphere]. 1997;**102**(D15):18865-18877

[33] Yonemura S, Sudo S, Tsuruta H, Kawashima S. Relations between light hydrocarbon, carbon monoxide, and carbon dioxide concentrations in the plume from the combustion of plant material in a furnace. Journal of Atmospheric Chemistry. 2002;**43**:1-19

[34] Yonemura S, Kawashima S. Concentrations of carbon gases and oxygen and their emission ratios from the combustion of rice hulls in a wind tunnel. Atmospheric Environment. 2007;**41**:1407-1416

[35] Hayashi K, Ono K, Kajiura M, Sudo S, Yonemura S, Fushimi A, et al. Trace gas and particle emissions from open burning of three cereal crop residues: Increase in residue moistness enhances emissions of carbon monoxide, methane, and particulate organic carbon. Atmospheric Environment. 2014;**95**: 36-44

[36] Burling IR, Yokelson RJ, Griffith DWT, Johnson TJ, Veres P, Roberts JM, et al. Laboratory measurements of trace gas emissions from biomass burning of fuel types from the southeastern and southwestern United States. Atmospheric Chemistry and Physics. 2010;**10**:11115-11130

[37] Friedli HR, Radke LF, Lu JY, Banic CM, Leaitch WR, PacPherson JI. Mercury emissions from burning of biomass from temperate north American forests: Laboratory and airborne measurements. Atmospheric Environment. 2003;**37**:253-267

[38] Obrist D, Moosmuller H, Schurmann R, Chen LWA, Kreidenwies SM. Particulate-phase and gaseous elemental mercury emissions during biomass combustion: Controlling factors and correlation with particulate matter emissions. Environmental Science and Technology. 2008;**42**(3):721-727

[39] Czapiewski KV, Czuba E, Huang L, Ernst D, Norman AL, Koppmann R, et al. Isotopic composition of non-methane hydrocarbons in emissions from biomass burning. Journal of Atmospheric Chemistry. 2002;**43**:45-60

[40] Friedli HR, Radke LF, Lu JY. Mercury in smoke from biomass fires. Geophysical Research Letters. 2001;**28**(17):3223-3226

[41] Barbosa PM, Stroppiana D, Gregoire JM, Miguel J, Pereira C. An assessment of vegetation fire in Africa (1981-1991): Burned areas, burned biomass, and atmospheric emissions. Global

Biogeochemical Cycles. 1999;**13**(4): 933-950

[42] Mouillot F, Narasimha A, Balkanski Y, Lamarque JF, Field CB. Global carbon emissions from biomass burning in the 20th century. Geophysical Research Letters. 2006;**33**:L01801

[43] Wiedinmyer C, Quayle B, Geron C, Belote A, McKenzie D, Zhang X, et al. Estimating emissions from fires in North America for air quality modeling. Atmospheric Environment. 2006;**40**: 3419-3432

[44] Ito A, Penner JE. Global estimates of biomass burning emissions based on satellite imagery for the year 2000. Journal of Geophysical Research. 2004; **109**:D14S05

[45] Akagi SK, Yokelson RJ, Wiedinmyer C, Alvarado MJ, Reid JS, Karl T, et al. Emission factors for open and domestic biomass burning for use in atmospheric models. Atmospheric Chemistry and Physics. 2011;**11**:4039-4072

[46] Andreae MO. Emissions of trace gases and aerosols from biomass burning – An updated assessment. Atmospheric Chemistry and Physics. 2019;**19**: 8523-8546

[47] Andreae MO, Merlet P. Emission of trace gasses and aerosols from biomass burning. Global Biogeochemical Cycles. 2001;**15**(4):955-966

[48] Koppmann R, von Czapiewski K, Reid JS. A review of biomass burning emissions, part I: Gaseous emissions of carbon monoxide, methane, volatile organic compounds, and nitrogen containing compounds. Atmospheric Chemistry and Physics Discussion. 2005; **5**:10455-10516

[49] Chen J, Li C, Ristovski Z, Milic A, Gu Y, Islam MS, et al. A review of biomass burning: Emissions and impacts on air quality, health and climate in China. Science of Total Environment. 2017;**579**:1000-1034

[50] Veiga MM, Meech JA, Oñante N. Mercury pollution from deforestation. Nature. 1994;**368**:816-817

[51] Wiedinmyer C, Friedli H. Mercury emission estimates from fires: An initial inventory for the United States. Environmental Science and Technology. 2007;**41**:8092-8098

[52] Biswas A, Blum JD, Klaue B, Keeler GJ. Release of mercury from Rocky Mountain forest fires. Global Geochemical Cycles. 2007;**21**:GB1002

[53] Engle MA, Gustin MS, Johnson DW, Murphy JF, Miller WW, Walker RF, et al. Mercury distribution in two Sierran forest and one desert sagebush steppe ecosystems and the effects of fire. Science of Total Environment. 2006;**367**: 222-233

[54] Brunke E-G, Labuschagne C. Gaseous mercury emissions from a fire in the cape peninsula, South Africa, during January 2000. Geophysical Research Letters. 2001;**28**(8):1483-1486

[55] Cinnirella S, Pirrone N. Spatial and temporal distributions of mercury emissions from forest fires in Mediterranean region and Russian federation. Atmospheric Environment. 2006;**40**:7346-7361

[56] Albernaz P, Michelazzo M, Fostier AH, Magarelli G, Santos JC, de Carvalho Jr. A. Mercury emissions from forest burning in southern Amazon. Geophysical Research Letters. 2010;**37**: L09809

[57] Sigler JM, Lee X, Munger W. Emission and long-range transport of

gaseous mercury from a large-scale Canadian boreal forest fire. Environmental Science and Technology. 2003;**37**:4343-4347

[58] Weiss-Penzias P, Jaffe D, Swartzendruber P, Hafner W, Chand D, Prestbo E. Quantifying Asian and biomass burning sources of mercury using the Hg/CO ratio in pollution plumes observed at the mount bachelor observatory. Atmospheric Environment. 2007;**41**:4366-4379

[59] Ebinghaus R, Slemr F, Brenninkmeijer CAM, van Velthoven P, Zahn A, Hermann M, et al. Emissions of gaseous mercury from biomass burning in South America in 2005 observed during CARIBIC flights. Geophysical Research Letters. 2007;**34**:L08813

[60] Friedli HR, Arelino FA, Cinnirella S Jr, Pirrone N. Chapter 8: Mercury emissions from global biomass burning: Spatial and temporal distribution. In: Pirrone N, Mason R, editors. Mercury Fate and Transport in the Global Atmosphere: Measurements, Models and Policy Implications. Interim Report of the UNEP Global Mercury Partnership Mercury Air Transport and Fate Research partnership area July 14. 2008. pp. 145-167

[61] Irei S. Isotopic characterization of gaseous mercury and particulate water-soluble organic carbon emitted from open grass field burning in Aso, Japan. Applied Sciences. 2022;**12**:109

[62] Irei S. Development of fast sampling and high recovery extraction method for stable isotope measurement of gaseous mercury. Applied Sciences. 2020;**10**:6691

[63] Du SH, Fang SC. Uptake of elemental mercury vapor by C3 and C4 species. Environmental and Experimental Botany. 1982;**22**(4):437-443

[64] Browne CL, Fang SC. Differential uptake of mercury vapor by Gramineous C3 and C4 plants. Plant Physiology. 1983;**72**:1040-1042

[65] Du SH, Fang SC. Catalase activity of C3 and C4 species and its relationship to mercury vapor uptake. Environmental and Experimental Botany. 1983;**23**(4): 347-353

[66] Hanson PJ, Lindberg SE, Tabberer TA, Owens JG, Kim K-H. Foliar exchange of mercury vapor: Evidence for compensation point. Water, Air, and Solid Pollution. 1995;**80**:373-382

[67] Lindberg SE. Forests and the global biogeochemical cycle of mercury: The importance of understanding air/vegetation exchange processes. In: Baeyens W, editor. Global and Regional Mercury Cycles: Sources, Fluxes, and Mass Balances. 1996. pp. 359-380

[68] Louis VLST, Rudd JWM, Kelly CA, Hall BD, Rolfhus KR, Scott KJ, et al. Importance of the forest canopy to fluxes of methyl mercury and total mercury to boreal ecosystems. Environmental Science and Technology. 2001;**35**:3089-3098

[69] Rea AW, Lindberg SE, Scherbatskoy T, Keeler GJ. Mercury accumulation in foliage over time in two northern mixed-hardwood forests. Water, Air, and Soil Pollution. 2002;**133**:49-67

[70] Ericksen JA, Gustin MS, Schorran DE, Johnson DW, Lindberg SE, Coleman JS. Accumulation of atmospheric mercury in forest foliage. Atmospheric Environment. 2003;**37**:1613-1622

[71] Millhollen AG, Gustin MS, Obrist D. Foliar mercury accumulation and exchange for three tree species. Environmental Science and Technology. 2006;**40**:6001-6006

[72] Fay L, Gustin M. Assessing the influence of different atmospheric and soil mercury concentrations on foliar mercury concentrations in a controlled environment. Water, Air, and Soil Pollution. 2007;**181**:373-384

[73] Friedli HR, Radke LF, Payne NJ, McRae DJ, Lynbam TJ, Blake TW. Mercry in vegetation and organic soil at an upland boreal forest site in prince albert national park. Saskatchewan. Journal of Geophysical Research. 2007; **112**:G01004

[74] Obrist D, Johnson DW, Edmons RL. Effects of vegetation type on mercury concentrations and pools in two adjacent coniferous and deciduous forests. Jounal of Plant Nutrition and Soil Science. 2012; **175**:68-77

[75] Yin R, Feng X, Meng B. Stable mercury isotope variation in rice plants (*Oryza sativa* L.) from the Wanshan mercury mining district, SW China. Environmental Science and Technology. 2023;**47**:2238-2245

[76] Fu X, Zhu W, Zhang H, Sommar J, Yu B, Yang X, et al. Depletion of atmospheric gaseous elemental mercury by plant uptake at Mt. Changbai, Northeast China. Atmospheric Chemistry and Physics. 2016;**16**:12861-12873

[77] Jiskra M, Sonke JE, Obrist D, Bieser J, Edinghaus R, Myhre CL, et al. A vegetation control on seasonal variations in global atmospheric mercury concentrations. Nature Geoscience. 2018;**11**:244-250

[78] Zhou J, Obrist D, Dastoor A, Jiskra M, Ryjkov A. Vegetation uptake of mercury and impacts on global cycling. Nature Reviews. 2021;**2**:269-284

[79] Demers JD, Blum JD, Zak DR. Mercury isotopes in a forested ecosystem: Implications for air-surface exchange dynamics and the global mercury cycle. Global Biogeochemical Cycles. 2013;**27**:222-238

[80] Yu B, Fu X, Yin R, Zhang H, Wang X, Lin CJ, et al. Isotopic compositions of atmospheric mercury in China: New evidence for sources and transformation processes in air and in vegetation. Environmental Science and Technology. 2016;**50**:9262-9269

[81] Zheng W, Obrist D, Weis D, Bergquist BA. Mercury isotope compositions across north American forests. Global Biogeochemical Cycles. 2016;**30**:1475-1492

[82] Jiskra M, Wiederhold JG, Skyllberg U, Kronberg RM, Hajdas I, Kretzschmar R. Mercury deposition and re-emission pathways in boreal forest soils investigated with Hg isotope signatures. Environmental Science and Technology. 2015;**49**:7188-7196

[83] Meng B, Li Y, Cui W, Jiang P, Liu G, Wang Y, et al. Tracing the uptake, transport, and fate of mercury in sawgrass (*Cladimu jamaicense*) in the Florida everglades using multi-isotope technique. Environmental Science and Technology. 2018;**52**(6):3384-3391

[84] Ministry of Environment, Japan. Japan's Mercury Emission Inventory (FY 2020) [Internet]. 2023. Available from: https://www.env.go.jp/content/000145228.pdf [Accessed: September 6, 2023]

[85] Cinnirella S, Pirrone N, Allegrini A, Guglietta D. Modeling mercury emissions from forest fires in the Mediterranean region. Environmental Fluid Mechanics. 2008;**8**:129-145

[86] Payra S, Sharma A, Verma S. Chapter 14: Application of remote

sensing to study forest fires. In: Singh AK, Tiwari S, editors. Atmospheric Remote Sensing Principles and Applications. 2022. pp. 239-260

[87] Kurmaz B, Baylk C, Abdikan S. Forest Fire Area Detection by Using Landsat-8 and Sentinel-2 Satellite Images: A Case Study in Mugla, Turkey. 2020. Available from: https://www.semanticscholar.org/paper/Forest-Fire-Area-Detection-by-Using-Landsat-8-and-A-Kurnaz-Bay%C4%B1k/655389305d5975f58ea07f2c0ed64121c13ca86b [Accessed: 2023]

[88] Hawbaker TJ, Vanderhoof MK, Schmidt GL, Beal Y-J, Picotte JJ, Takacs JD, et al. The Landsat burned area algorithm and products for the conterminous United States. Remote Sensing of Environment. 2020;**244**: 111801

[89] Ying L, Shen Z, Yang M, Piao S. Wildfire detection probability of MODIS fire products under the constraint of environmental factors: A study based on confirmed ground wildfire records. Remote Sensing. 2019;**11**(24):3031

[90] Ward DE, Radke LF. Chapter 4: Emissions measurements from vegetation fires: A comparative evaluation of methods and results. In: Crutzen PJ, Goldammer JG, editors. The Ecological, Atmospheric, and Climatic Importance of Vegetation Fires. City: John Wiley & Sons Ltd.; 1993. pp. 53-76. ISBN 0-471-93604-9

[91] Kumamoto Prefecture. Basic Research for Aso Grassland Regeneration Plan [Internet]. 2022. Available from: https://www.pref.kumamoto.jp/uploaded/life/181104_443400_misc.pdf [Accessed: 2023]

[92] Fu X, Heimbürger L-E, Sonke JE. Collection of atmospheric gaseous mercury for stable isotope analysis using iodine- and chlorine-impregnated activated carbon traps. Journal of Analytical Atomic Spectrometry. 2014; **29**:841

[93] Yamakawa A, Takami A, Takeda Y, Kato S, Kajii Y. Emerging investigator series: Investigation of mercury emission sources using Hg isotopic compositions of atmospheric mercury at the cape Hedo atmosphere and aerosol monitoring station (CHAAMS). Japan. Environmental Science and Processing Impacts. 2019;**21**:809-818

[94] Inomata H, Qong M, Suzuki Y, Fukuma M. Temporal variation of burned area of grassland in Aso – Analysis of landsat imagery (in Japanese). Japanese Journal of Grassland Science. 2006;**51**(4):341-347

[95] Bilbao B, Medina E. Types of grassland fires and nitrogen volatilization in tropical savannas of Calabozo, Venezuela. Chapter 55 in Book Biomass Burning and Global Change Volume 2 Biomass Burning in South America, Southeast Asia, and Temperate and Boreal Ecosystems, and the Oil Fires of Kuwait

[96] Ohta Y. Area of controlled burning of grassland on Akiyoshi-dai plateau. Bulletin of the Akiyoshi-dai Museum of Natural History. 2018;**53**:23-26. (in Japanese)

[97] Blum JD, Bergquist BA. Reporting of variations in the natural isotopic composition of mercury. Analytical and Bioanalytical Chemistry. 2007;**388**: 353-359

Section 2

Grasslands Development Initiatives

Chapter 5

Integrating Livestock and Crop Systems for Enhanced Productivity and Grassland Conservation in Developing Countries

Rahmathulla Mohamed Nikzaad and Noordeen Nusrathali

Abstract

Many conventional farming approaches in developing nations segregate livestock and crop production, ignoring the synergistic advantages and sustainable land management possibilities that may be gained by combining the two. In order to increase agricultural output and foster grassland conservation, this chapter investigates the idea of merging livestock and crop systems. By highlighting the advantages and challenges of the approach, this chapter draws attention to the potential benefits of integration, including enhanced soil fertility, efficient resource use, increased productivity, and better protection of grassland ecosystems. The importance and viability of a variety of integrated agricultural methods, including agro-pastoral, mixed, and silvopastoral systems, in a variety of geographical settings, are explored. The purpose of this chapter is to educate policymakers, academics, and practitioners on the need of integrating livestock and crop production for achieving long-term agricultural sustainability in low-income nations.

Keywords: grass land, conservation, climate change, livestock crop, developing countries

1. Introduction

1.1 Background

Agriculture is the lifeline of many developing nations, sustaining millions of people by providing them with employment and an essential income [1]. However, poor production, soil erosion, and inefficient use of resources are common problems for conventional farming in these areas. The need for sustainable agriculture techniques is becoming more urgent as the world's population is expected to rise to roughly 9.7 billion by 2050 and climate change offers greater dangers [2]. Innovative solutions are needed to meet the challenge of increasing food production while reducing environmental consequences and protecting scarce resources. Among them, integrating livestock and agricultural systems stands out as a method with the potential to help developing nations overcome obstacles and advance sustainable land management [3, 4]. This progressive kind of farming takes into account all aspects of food production, rather than just the crops and cattle [5]. Instead, it promotes cooperation between these two agricultural pillars in order to maximize gains in production, grassland ecosystem preservation, and rural community prosperity [4].

Instead of keeping crop farming and animal farming separate, as is common in traditional farming approaches, integrated farming takes a more holistic and creative approach [6]. To solve the problems of contemporary agriculture, it takes into account the symbiotic relationship between crop production and animal husbandry [6]. The goal of this method is to increase agricultural output while also protecting grassland ecosystems and improving the quality of life in rural areas [4]. Traditional and indigenous agricultural traditions have long acknowledged the advantages of reciprocal interactions between crops and animals, therefore the idea of integrating livestock and crop systems is not new [7]. These time-tested methods have always known the need of coordinating the ecological roles of crops and livestock within a single system [8]. Industrialization and the Green Revolution, which prioritized specialized and intensive farming, led to the widespread replacement of these integrated techniques with more standardized and simpler ways [3, 4].

Recent years, however, have seen a resurgence in interest in these integrated systems and their promotion as a long-term answer to the problems plaguing contemporary agriculture [9–11]. Integrating livestock and agricultural systems has the potential to overcome many of the drawbacks of conventional farming [6]. New methods that improve resource use and nitrogen cycling might slow the decline of agricultural productivity and soil quality [12]. Including livestock in agricultural systems has the potential to help farmers due to the nutrient-rich manure created by animals [13]. Cattle may be given crop scraps in return, which is good for both parties [14]. Furthermore, this approach offers the possibility of safeguarding grassland ecosystems and preserving biodiversity [7]. Protecting grassland biodiversity, halting soil erosion, and improving ecosystem health may all be possible via the use of sustainable grazing practices within integrated systems [15].

Integrating livestock and crop production allows for more diversity and stability. Having many streams of income may help farmers weather economic storms and improve their financial stability [16]. Having additional food alternatives is one way in which a diverse agricultural system contributes to food security [17]. Although there is much to be gained by implementing integrated agricultural systems, there are a number of challenges that must be overcome before they can be implemented on a large scale. Raising awareness and educating the public may help increase the adoption of integrated systems [18]. It is critical to give farmers with knowledge, training, and extension services to aid in their transition to more sustainable and integrated practices [18].

Agricultural techniques may also be affected by sociocultural issues. Encouraging farmers to adopt new methods requires an appreciation for, and an understanding of, traditional knowledge and beliefs [19]. Creating an enabling environment for the broad adoption of integrated agricultural methods requires supportive policies, incentives, and institutional structures. The promise of integrated systems must be acknowledged by policymakers, who must then enact enabling legislation to encourage farmers to embrace this sustainable method [20]. Technology and infrastructure play critical roles in making integrated systems more practical. Integrated agricultural approaches are more likely to be adopted successfully when farmers have access to better seed varieties, animal types, and market connections [21].

DOI: http://dx.doi.org/10.5772/intechopen.113109

Consequently, the issues experienced by conventional agricultural methods in poor countries may be mitigated by the integration of livestock and crop systems [22]. Farmers may increase output, protect ecosystems, and better the lives of people in rural areas by understanding the interdependence of crops and animals and using this knowledge to their advantage [3, 4]. Integrated livestock and crop systems may lead the way towards a more resilient, productive, and environmentally aware agricultural future in developing nations via joint efforts and a commitment to sustainable practices [10]. Food security, economic prosperity, and ecological preservation may all be ensured for future generations by adopting and promoting integrated farming techniques in developing nations [23].

The primary objectives of this chapter are to perform a thorough research and analysis of traditional agricultural methods in developing countries and to emphasize the limitations and hazards that these practices provide to sustainable agriculture and food security. In order to stress the need of combining livestock and agricultural systems as a sustainable and successful approach to land management in developing countries, this chapter highlights potential benefits such as higher productivity, improved soil health, and grassland conservation. Different types of integrated farming will be examined, including agro-pastoral systems, mixed farming, and silvopastoral systems, and their relative relevance and feasibility in different types of environments. This chapter will also cover the issues and methods necessary to implement integrated livestock and agricultural systems, such as land management practices, animal waste management, and the incorporation of technology. Integration's social and economic implications, such as its impact on rural residents' incomes, food security, and climate change adaptation preparedness, will also be examined. The purpose of this chapter is to contribute to the existing body of knowledge on sustainable agricultural approaches in developing countries by assisting policymakers, researchers, and farmers in supporting integrated livestock and crop systems for greater productivity and grassland conservation.

2. Advantages and challenges of integrating livestock and crop systems

2.1 Advantages of integrating livestock and crop systems

2.1.1 Enhanced nutrient cycling

One of the primary advantages of integrating livestock and crop systems is the efficient nutrient cycling it facilitates [12]. Livestock produces organic matter in the form of manure, which can be utilized as natural fertilizers for crops [24]. This reduces the dependence on synthetic chemical fertilizers, closing nutrient loops, and promoting sustainable soil health and fertility [24]. In return, crop residues and by-products can serve as feed and forage for livestock, minimizing waste and maximizing resource utilization [25] (**Figure 1**).

2.1.2 Improved soil health

The integration of livestock and crop systems contributes to improved soil health [27]. Livestock grazing can help break up compacted soils, improve soil aeration, and stimulate biological activity [28]. Additionally, the organic matter from livestock manure enriches the soil, enhancing its water-holding capacity and nutrient content. Healthy soils support better crop growth and resilience to environmental stresses [27].

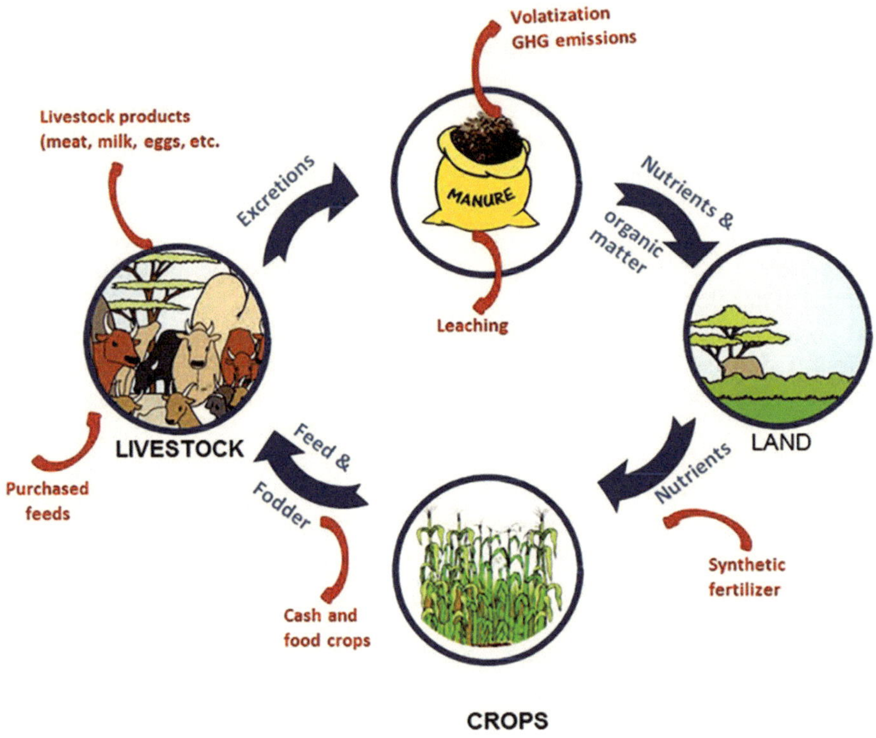

Figure 1.
Nutrient cycle in crop-livestock systems [26].

2.1.3 Grassland conservation

Sustainable grazing practices within integrated systems can play a vital role in grassland conservation [15]. Properly managed rotational grazing allows for rest and recovery periods for pastures, preventing overgrazing and maintaining biodiversity [29]. By preserving natural habitats and avoiding grassland conversion, integrated systems contribute to the conservation of grassland ecosystems and the wildlife they support [30].

2.1.4 Diversification of income streams

Integrating livestock and crop systems offers farmers the opportunity to diversify their income streams. Relying on both crops and livestock provides multiple sources of revenue, reducing the vulnerability to market fluctuations and ensuring economic stability for rural communities [4].

2.1.5 Resilience to climate change

Integrated systems can enhance resilience to climate change impacts. Diversification of production and income sources, as well as improved soil health, can make farms more adaptable to changing climatic conditions, such as altered rainfall patterns and extreme weather events [31].

2.1.6 Efficient resource utilization

By combining crops and livestock, farmers can optimize resource utilization. For instance, livestock can graze on crop residues and cover crops, reducing the need for feed supplements. In turn, crop residues and by-products can be used as supplementary feed for livestock [32] (**Figure 2**).

2.2 Challenges of integrating livestock and crop systems

2.2.1 Knowledge and awareness

The requirement for knowledge and awareness among farmers is a major obstacle to integrating livestock and agricultural systems. Adopting new integrated techniques in farming involves access to knowledge, training, and extension services [33, 34]. Many traditional agricultural practices have been handed down from generation to generation.

2.2.2 Socio-cultural factors

Agricultural methods may be affected by cultural and social norms. Engaging with local people and respecting their traditional knowledge is vital since introducing new techniques may be met with hostility if they disagree with deeply ingrained practices or beliefs [6].

Figure 2.
Advantages of crop live integration.

2.2.3 Policy and institutional support

The adoption of integrated agricultural methods relies heavily on enabling policies and institutional structures. To fully realize integration's potential, governments must acknowledge its value and reward it, empowering farmers with the means they need to implement and maintain such systems [4].

2.2.4 Technology and infrastructure

It might be difficult to develop integrated livestock and agricultural systems in areas with limited access to necessary technologies and infrastructure. Facilitating the adoption of integrated techniques requires better access to improved crops, animal breeds, and market links [35].

2.2.5 Land management and space constraints

In order to meet the demands of both crops and cattle, land must be managed carefully in an integrated system. Proper rotational grazing may be difficult in areas with limited acreage, necessitating creative land use planning to maximize output [36].

2.2.6 Animal waste management

Animal waste must be properly managed to avoid pollution and protect public health. When incorporating animals into agricultural systems, it is crucial to develop effective and sustainable waste management procedures [37].

3. Integrated farming approaches

Integrated farming approaches encompass various systems that combine crop cultivation and livestock rearing in a synergistic manner [38]. These approaches aim to optimize resource utilization, enhance productivity, and promote sustainable land management [38]. Each integrated farming system adapts to the specific ecological, social, and economic context of a region, offering flexibility in implementation [39]. Some of the key integrated farming approaches include:

4. Agro-pastoral systems

Agro-pastoral systems integrate crop production and livestock grazing in the same area [40]. Livestock, such as cattle, sheep, or goats, are allowed to graze on fallow or harvested crop fields, consuming crop residues and weeds [40]. The manure from livestock returns nutrients to the soil, benefiting subsequent crop growth. This system fosters a balanced nutrient cycle, conserves grassland ecosystems, and promotes biodiversity by alternating between grazing and crop production [12].

4.1 Mixed farming

Mixed farming involves combining both crops and livestock on the same farm, but in distinct areas [41]. Farmers allocate specific plots for crop cultivation and maintain separate areas for livestock grazing or raising [41]. The integration occurs at the farm level, with complementary interactions between crops and livestock. For example, crop residues and by-products can be used as animal feed, while manure enriches the soil for crop cultivation [42].

4.2 Silvopastoral systems

Silvopastoral systems combine trees or agroforestry components with livestock grazing. In these systems, farmers plant trees or shrubs on grazing lands, providing shade and shelter for livestock [43]. The trees also contribute to soil improvement through their root systems and enhance biodiversity. Livestock, in turn, contribute to tree management by pruning and seed dispersal, aiding in natural regeneration [44]. Silvopastoral systems promote carbon sequestration, mitigate climate change impacts, and improve the overall ecological resilience of the farming landscape [45].

4.3 Integrated aquaculture-agriculture systems

This approach combines fish or aquatic organism farming with crop production. Farmers integrate fish ponds or aquaculture tanks within their agricultural fields or in proximity to them [46]. Fish wastes serve as natural fertilizers for crops, while the crops provide shade and natural food sources for the fish. This system enhances overall farm productivity and diversifies income streams [47].

4.4 Integrated crop-livestock-forest systems

This system integrates crop cultivation, livestock rearing, and forest components within the same agricultural landscape. Farmers cultivate crops, raise livestock, and maintain forests or agroforestry areas. The system maximizes resource utilization, conserves biodiversity, and offers economic and environmental benefits [48].

Integrated farming approaches offer numerous advantages for enhancing productivity and promoting sustainable land management in developing countries. By optimizing resource utilization, conserving biodiversity, and promoting resilience to climate change, these systems pave the way towards a more sustainable and prosperous agricultural future [4]. Addressing challenges through knowledge dissemination, institutional support, and socio-cultural awareness will be critical in facilitating the widespread adoption of integrated farming approaches [49].

5. Socio-economic implications

5.1 Rural livelihoods

Communities who depend substantially on agriculture as a source of income may feel the greatest effects of integrated agricultural systems [16]. Increasing economic stability and decreasing reliance on changes in any one sector [16] is possible for rural communities by combining several agricultural activities, such as crop agriculture, livestock husbandry, aquaculture, and agroforestry.

5.1.1 Diversification of income streams and poverty alleviation

Monoculture, in which farmers only engage in one kind of farming, is common in conventional agriculture. This may be dangerous if the crop fails because of things like pests, illnesses, or bad weather [50]. However, farmers that use integrated systems might increase their revenue by diversifying their sources of income. Farmers may multitask by tending to many fields, herds, and tanks at once [4]. By reducing the possibility of total income loss and spreading the risk, diversification may help relieve poverty in rural areas [51].

5.1.2 Enhancing livelihood resilience through integrated practices

The resilience of rural lifestyles is boosted by integrated systems as well. Farmers who engage in a variety of endeavors are better able to respond to unexpected changes. They have many revenue streams to fall back on in the event of crop failure [52]. Furthermore, sustainable and regenerative methods are commonly emphasized in integrated systems, which assist conserve natural resources and sustain land productivity over the long term, ensuring future generations' ability to make a living [52].

5.2 Food security

5.2.1 Contributing to food production and availability

Integrated systems contribute significantly to food production and availability. By combining different agricultural practices, farmers can optimize land use and resource allocation [33]. For instance, they can use animal manure as fertilizers for crops, and crop residues can be fed to livestock, creating a closed-loop system that maximizes productivity [33]. This efficient use of resources helps increase food production and availability, reducing the risk of food shortages in rural areas.

5.2.2 Strengthening food security in vulnerable regions

Integrated systems can play a crucial role in enhancing food security in vulnerable regions, particularly those prone to environmental and climate-related challenges [53]. By adopting climate-resilient practices, farmers can mitigate the impact of adverse conditions like droughts, floods, or extreme temperatures on their crops and livestock. Moreover, diversification of crops and livestock provides a buffer against food crises caused by localized crop failures or diseases [54].

5.2.3 Promoting dietary diversity and nutrition

Through integrated systems, farmers can grow a diverse range of crops and rear different types of livestock. This diversity extends to the local food supply and promotes dietary diversity among rural communities [55]. Consuming a varied diet with a mix of fruits, vegetables, grains, and animal products can significantly improve nutrition and overall health, reducing the prevalence of malnutrition and diet-related health issues [55].

5.3 Climate change resilience

5.3.1 How integrated systems can enhance resilience to climate change?

Climate change poses a significant threat to agricultural productivity, with increased occurrences of extreme weather events and shifting climatic patterns [56]. Integrated systems are more resilient to these changes because they offer flexibility and adaptability. By combining various farming practices, farmers can adjust their operations in response to changing conditions. For instance, during droughts, they can focus on drought-resistant crops and reduce water-intensive practices [57].

5.3.2 Mitigating climate-related risks through diversified production

Integrated systems also mitigate climate-related risks by reducing dependence on a single crop or livestock species. If one element of the system is affected by a climate-related hazard, other components can still provide income and sustenance [58]. This reduces the vulnerability of farmers to the adverse effects of climate change, such as crop failures or livestock losses [33].

5.3.3 Carbon sequestration potential of integrated farming

One important aspect of climate change resilience is the potential to sequester carbon dioxide from the atmosphere [59]. Integrated farming practices, especially agroforestry and cover cropping, can enhance carbon sequestration in the soil and vegetation. Trees and cover crops capture carbon, mitigating greenhouse gas emissions and contributing to climate change mitigation efforts [60].

6. Case studies

6.1 Case studies from Africa

Examples of Successful Integration in African Countries:

Case Study 1: In Zambiya, a farming community successfully implemented an integrated farming system that combined crop cultivation with fish farming (aquaponics). They used the nutrient-rich water from fish ponds to fertilize the crops, while the crops filtered the water for the fish. This integration led to increased crop yields and fish production, improving food security and providing additional income streams for the community [61].

Case Study 2: In Ethiopia, a project focused on integrating agroforestry with traditional crop cultivation. Farmers planted trees alongside their crops, providing shade and windbreaks, which reduced soil erosion and water loss. The trees also contributed fruits, nuts, and timber, adding to the farmers' income. This sustainable approach improved soil fertility, enhanced climate resilience, and empowered the local communities economically [62].

6.2 Case studies from Asia

Integrating Livestock and Crop Systems in Asian Nations:

Case Study 1: In India, a dairy cooperative introduced an integrated model where crop residues and by-products were utilized as animal feed. This reduced the pressure on natural resources and minimized waste while increasing milk productivity. The cooperative also promoted organic farming practices, enhancing the overall sustainability of the system [63].

Case Study 2: In Vietnam, the combination of rice farming with duck rearing has proven successful. Ducks forage in rice paddies, consuming insects and weeds, which reduces the need for chemical pesticides and manual weeding. The ducks also provide an additional income source through egg and meat production [64].

6.3 Case studies from Latin America

Experiences of Integrating Systems in Latin American Countries:

Case Study 1: In Brazil, a cooperative of small-scale farmers adopted an integrated system that combined cattle ranching with silvopastoral practices. They planted trees and forage crops in pasture areas, providing shade and nutritious fodder for the cattle. This approach reduced deforestation, improved livestock health, and increased the overall productivity of the land [65].

Case Study 2: In Colombia, coffee farmers integrated beekeeping into their coffee plantations. Bees enhanced coffee pollination, leading to higher coffee yields. Additionally, the farmers could sell honey and beeswax, diversifying their income sources and promoting biodiversity [66].

6.4 Challenges unique to the region and potential solutions

Land Fragmentation: Many Asian countries face land fragmentation due to inheritance laws. Integrated systems might require larger land areas, which can be challenging to assemble. Land consolidation programs or community-based land-sharing arrangements could be potential solutions [67].

Water Management: In regions with water scarcity, managing water resources becomes critical for integrated systems. Implementing water-saving irrigation techniques, rainwater harvesting, and efficient water use can address this challenge [68].

Traditional Practices: Convincing farmers to shift from traditional practices to integrated systems can be challenging. Demonstrating the economic and environmental benefits through pilot projects and on-farm trials can help build confidence and encourage adoption [69].

7. Policy and institutional support

7.1 Policy frameworks

Analyzing Existing Agricultural Policies and their Impact on Integration:

Before promoting integrated farming, it is crucial to assess the existing agricultural policies to understand their impact on integration. Some policies may inadvertently hinder the adoption of integrated systems due to their focus on monoculture or specific agricultural sectors [70]. Additionally, subsidies and incentives may favor conventional farming practices, making it challenging for farmers to transition to integrated approaches [71]. An in-depth analysis can help identify gaps and areas where policy adjustments are needed to encourage integration.

7.2 Policy recommendations to support integrated farming

Incentives for Diversification: Governments can introduce financial incentives, such as subsidies or tax breaks, to encourage farmers to adopt integrated farming practices. These incentives could target crop-livestock integration, agroforestry, aquaponics, or other sustainable practices that enhance resilience and productivity [72].

Research and Development Funding: Governments should allocate funds for research and development specific to integrated systems. This investment can support the creation of best practices, improved technologies, and knowledge dissemination to farmers through extension services [73].

Land Tenure Security: Secure land tenure is essential for farmers to invest in integrated systems, as many practices involve long-term planning. Governments can provide land tenure guarantees and clarify land rights to instill confidence in farmers to adopt integrated approaches [74].

Market Access and Value Chains: Strengthening market linkages for integrated farming products can incentivize farmers to adopt such practices. Governments can facilitate access to markets, promote fair trade practices, and support value addition initiatives to increase the profitability of integrated systems [75].

Sustainable Agriculture Standards: Integrate sustainable agriculture standards into policy frameworks to encourage the adoption of ecologically friendly and socially responsible practices. These standards can guide farmers towards integrated approaches and promote environmentally friendly production [76].

7.3 The way forward for policymakers, researchers, and farmers

The role of policymakers in facilitating integrated farming is vital [77]. They need to give sustainable farming more weight in legislative frameworks, incentivize the use of integrated systems, fund studies on their efficacy, and bolster extension services for farmers [77].

Scientists should keep looking for new ways to improve the efficiency and scalability of integrated agricultural methods. Context-specific and implementable solutions may be found via collaborative research initiatives including stakeholders from academics, governments, and agricultural communities [73].

The use of integrated systems by farmers should be promoted and supported. Farmers may make a smooth shift to sustainable methods with the support of training programs, extension services, and access to capital and markets [78].

Food security, climate change, and poverty may all be combated via sustainable agriculture if livestock and agricultural systems are integrated. Agriculture and rural communities may have a more secure and sustainable future if governments, academics, and farmers all adopt this strategy.

7.4 Institutional capacity building

Strengthening Institutions for Promoting Integrated Systems:

Promoting integrated farming in developing countries requires a tailored approach that addresses the unique challenges and opportunities in these regions. Several studies and reports highlight the importance of strengthening institutions, providing training and extension services, and creating supportive policy frameworks to foster integrated farming practices.

1. **Strengthening Institutions**: The Food and Agriculture Organization (FAO) emphasizes the significance of strengthening agricultural institutions in developing countries to promote integrated farming. Creating specialized departments or units within existing agricultural agencies can enhance the focus on integrated systems. These units can provide technical support, facilitate research, and coordinate with relevant stakeholders [79].

2. **Training and Extension Services**: The International Fund for Agricultural Development (IFAD) recognizes the importance of training and extension services for smallholder farmers in developing countries. Organizing workshops, training programs, and demonstration farms can effectively showcase the benefits of integrated farming and teach farmers best practices [80].

3. **Farmer-to-Farmer Knowledge Exchange Programs**: Research conducted by the International Journal of Agricultural Extension highlights the effectiveness of farmer-to-farmer knowledge exchange programs in developing countries. Such programs enable successful adopters of integrated farming systems to share their experiences and knowledge with neighboring farmers, leading to higher adoption rates and stronger community support [81].

4. **Leveraging Digital Technologies**: In the context of developing countries, the World Bank emphasizes the potential of digital technologies and online platforms to disseminate agricultural information. Access to information through digital means can empower farmers in remote areas with knowledge about integrated farming practices [82].

5. **Supportive Policy Frameworks**: The United Nations Development Programme (UNDP) stresses the importance of supportive policy frameworks for integrated farming in developing countries. Governments need to analyze existing policies, make necessary adjustments, and introduce new incentives to create an enabling environment for farmers to adopt integrated approaches [83].

By implementing these strategies in a developing country context, governments and relevant stakeholders can foster the adoption of integrated farming practices, leading to increased agricultural productivity, improved environmental sustainability, and enhanced livelihoods for smallholder farmers.

8. Future perspectives and conclusions

8.1 Future prospects of integration

8.1.1 Potential for scaling up integrated systems in developing countries

There is great potential for expanding integrated agricultural systems in underdeveloped regions. The strain on agricultural systems to increase food production while being environmentally friendly will only increase as the world's population rises. Meeting this need while also resolving environmental and social issues is possible via the use of integrated techniques. Food security, poverty reduction, and rural development are all aided by integrated systems because they encourage diversity, resource efficiency, and resilience [4].

Governments, international organizations, and the corporate sector are more likely to support sustainable agriculture if public opinion shifts in its favor. Knowledge sharing and the use of lessons gained from regional case studies may also help with the scaling up of integrated systems [84].

8.1.2 Integrating new technologies and ideas into agriculture

Recent technological developments have been essential in spreading the word about integrated farming. Farms may be monitored and managed more effectively with the use of precision agricultural technology including remote sensing, drones, and sensor-based systems. These advancements allow for more precise input distribution, which in turn leads to less waste and higher output [85].

In addition, best practices, weather predictions, market pricing, and financial services are all more easily accessible to farmers thanks to digital platforms and mobile apps. With this knowledge, smallholder farmers will be better able to manage their resources and make educated agricultural decisions [86].

8.2 Conclusion

Integrating livestock and crop systems holds significant promise for advancing grassland conservation and environmental sustainability in developing nations. This approach offers multifaceted benefits, including heightened resource efficiency, amplified production, resilience to climate change, and preservation of biodiversity. By capitalizing on the synergies between livestock and crop production, farmers can optimize nutrient recycling, minimize waste, and reduce reliance on external inputs, resulting in increased yields, efficient land use, and sustainable resource management. This integrated model not only bolsters income stability for farmers by diversifying revenue sources but also fortifies resilience against climatic fluctuations and fosters adaptable production strategies. Overcoming hurdles like limited knowledge and institutional support remains essential for widespread adoption. Empowering farmers through educational initiatives, knowledge-sharing platforms, and tailored policies can bolster their confidence and competence in implementing integrated methods. Policymakers must champion sustainable practices, foster supportive frameworks, and enhance institutions to facilitate this transition and ensure equitable access to capital, inputs, and markets. In this collaborative endeavor, integrated livestock and agricultural systems emerge as a transformative solution to address pressing challenges, steering developing nations towards a sustainable and ecologically harmonious agricultural future.

Author details

Rahmathulla Mohamed Nikzaad* and Noordeen Nusrathali
Faculty of Technology, Department of Biosystems Technology, South Eastern University of Sri Lanka, Sri Lanka

*Address all correspondence to: mnikzaad@seu.ac.lk

References

[1] Jana NC, Singh A, Singh RB, editors. Livelihood Enhancement Through Agriculture, Tourism and Health. Singapore: Springer; 2022

[2] Kumar L, Chhogyel N, Gopalakrishnan T, Hasan MK, Jayasinghe SL, Kariyawasam CS, et al. Climate change and future of agri-food production. In: Future Foods. United States: Academic Press; 2022. pp. 49-79

[3] Leite FFGD, Nobrega GN, Baumgärtner LC, Alecrim FB, da Silveira JG, Cordeiro RC, et al. Greenhouse gas emissions and carbon sequestration associated with integrated crop-livestock-forestry (ICLF) systems. Environmental Reviews. 2023;**2023**:2-18

[4] Sekaran U, Lai L, Ussiri DA, Kumar S, Clay S. Role of integrated crop-livestock systems in improving agriculture production and addressing food security–a review. Journal of Agriculture and Food Research. 2021;**5**:100190

[5] Franzluebbers AJ, Sulc RM, Russelle MP. Opportunities and challenges for integrating North-American crop and livestock systems. Grassland Productivity and Ecosystem Services. 2011;**2011**:208-218

[6] Muhie SH. Novel approaches and practices to sustainable agriculture. Journal of Agriculture and Food Research. 2022;**2022**:100446

[7] Kremsa VŠ. Sustainable management of agricultural resources (agricultural crops and animals). In: Sustainable Resource Management. United States: Elsevier; 2021. pp. 99-145

[8] Capra F. Sustainable living, ecological literacy, and the breath of life. Canadian Journal of Environmental Education (CJEE). 2007;**2007**:9-18

[9] Horlings LG, Marsden TK. Towards the real green revolution? Exploring the conceptual dimensions of a new ecological modernisation of agriculture that could 'feed the world'. Global Environmental Change. 2011;**21**(2):441-452

[10] Lykogianni M, Bempelou E, Karamaouna F, Aliferis KA. Do pesticides promote or hinder sustainability in agriculture? The challenge of sustainable use of pesticides in modern agriculture. Science of the Total Environment. 2021;**795**:148625

[11] Trivedi P, Mattupalli C, Eversole K, Leach JE. Enabling sustainable agriculture through understanding and enhancement of microbiomes. New Phytologist. 2021;**230**(6):2129-2147

[12] Rao KPC, Verchot LV, Laarman J. Adaptation to climate change through sustainable management and development of agroforestry systems. Journal of SAT Agricultural Research. 2007;**4**(1):1-30

[13] Tarawali SA, Larbi A, Fernández-Rivera S, Bationo A. The contribution of livestock to soil fertility. Sustaining Soil Fertility in West Africa. 2001;**58**:281-304

[14] Gupta V, Rai PK, Risam KS. Integrated crop-livestock farming systems: A strategy for resource conservation and environmental sustainability. Indian Research Journal of Extension Education. 2012;**2**:49-54

[15] Dong S, Shang Z, Gao J, Boone RB. Enhancing sustainability of grassland

ecosystems through ecological restoration and grazing management in an era of climate change on Qinghai-Tibetan Plateau. Agriculture, Ecosystems & Environment. 2020;**287**:106684

[16] Sneessens I, Sauvée L, Randrianasolo-Rakotobe H, Ingrand S. A framework to assess the economic vulnerability of farming systems: Application to mixed crop-livestock systems. Agricultural Systems. 2019;**176**:102658

[17] Mustafa MA, Mayes S, Massawe F. Crop diversification through a wider use of underutilised crops: A strategy to ensure food and nutrition security in the face of climate change. Sustainable Solutions for Food Security: Combating Climate Change by Adaptation. 2019;**2019**:125-149

[18] Rose DC, Sutherland WJ, Barnes AP, Borthwick F, Ffoulkes C, Hall C, et al. Integrated farm management for sustainable agriculture: Lessons for knowledge exchange and policy. Land Use Policy. 2019;**81**:834-842

[19] Ilieva RT, Cohen N, Israel M, Specht K, Fox-Kämper R, Fargue-Lelièvre A, et al. The socio-cultural benefits of urban agriculture: A review of the literature. Land. 2022;**11**(5):622

[20] OECD. Adoption of Technologies for Sustainable Farming Systems. United States: OECD; 2001. p. 149

[21] Swaminathan MS. Can science and technology feed the world in 2025? Field Crops Research. 2007;**104**(1-3):3-9

[22] Kirchmann H, Thorvaldsson G. Challenging targets for future agriculture. European Journal of Agronomy. 2000;**12**(3-4):145-161

[23] Scialabba N. Factors influencing organic agriculture policies with a focus on developing countries. In: IFOAM 2000 Scientific Conference. Basel, Switzerland; 2000. pp. 28-31

[24] Vanlauwe B, Wendt J, Diels J. Combined application of organic matter and fertilizer. Sustaining Soil Fertility in West Africa. 2001;**58**:247-279

[25] Jayasuriya MCN. Use of crop residues and agro-industrial by-products in ruminant production systems in developing countries. BSAP Occasional Publication. 1993;**16**:47-55

[26] Ndambi OA, Pelster DE, Owino JO, De Buisonje F, Vellinga T. Manure management practices and policies in sub-Saharan Africa: Implications on manure quality as a fertilizer. Frontiers in Sustainable Food Systems. 2019;**3**:29

[27] Kumari VV, Balloli SS, Ramana DBV, Kumar M, Maruthi V, Prabhakar M, et al. Crop and livestock productivity, soil health improvement and insect dynamics: Impact of different fodder-based cropping systems in a rainfed region of India. Agricultural Systems. 2023;**208**:103646

[28] Sheoran V, Sheoran AS, Poonia P. Soil reclamation of abandoned mine land by revegetation: A review. International Journal of Soil, Sediment and Water. 2010;**3**(2):13

[29] Grandin T. Grazing cattle, sheep, and goats are important parts of a sustainable agricultural future. Animals. 2022;**12**(16):2092

[30] Sanderson MA, Archer D, Hendrickson J, Kronberg S, Liebig M, Nichols K, et al. Diversification and ecosystem services for conservation agriculture: Outcomes from pastures and integrated crop–livestock systems.

Renewable Agriculture and Food Systems. 2013;**28**(2):129-144

[31] Altieri MA, Nicholls CI, Henao A, Lana MA. Agroecology and the design of climate change-resilient farming systems. Agronomy for Sustainable Development. 2015;**35**(3):869-890

[32] Oosting S, van der Lee J, Verdegem M, de Vries M, Vernooij A, Bonilla-Cedrez C, et al. Farmed animal production in tropical circular food systems. Food Security. 2022;**14**(1):273-292

[33] Gomez-Zavaglia A, Mejuto JC, Simal-Gandara J. Mitigation of emerging implications of climate change on food production systems. Food Research International. 2020;**134**:109256

[34] Bhatt R, Kaur R, Ghosh A. Strategies to practice climate-smart agriculture to improve the livelihoods under the rice-wheat cropping system in South Asia. Sustainable Management of Soil and Environment. 2019;**2019**:29-71

[35] Garrett RD, Niles MT, Gil JD, Gaudin A, Chaplin-Kramer R, Assmann A, et al. Social and ecological analysis of commercial integrated crop livestock systems: Current knowledge and remaining uncertainty. Agricultural Systems. 2017;**155**:136-146

[36] Tiemann T, Douxchamps S. Opportunities and challenges for integrated smallholder farming systems to improve soil nutrient management in Southeast Asia. World Development Sustainability. 2023;**2023**:100080

[37] Hajam YA, Kumar R, Kumar A. Environmental waste management strategies and Vermi transformation for sustainable development. Environmental Challenges. 2023;**2023**:100747

[38] Gurjar GN, Swami S. Integrated farming systems of valley regions: Food and nutritional security. International Journal of Chemical Studies. 2019;**7**(2):773-778

[39] Bonaudo T, Bendahan AB, Sabatier R, Ryschawy J, Bellon S, Leger F, et al. Agroecological principles for the redesign of integrated crop–livestock systems. European Journal of Agronomy. 2014;**57**:43-51

[40] Ayantunde AA, De Leeuw J, Turner MD, Said M. Challenges of assessing the sustainability of (agro)-pastoral systems. Livestock Science. 2011;**139**(1-2):30-43

[41] Devendra C, Thomas D. Crop–animal interactions in mixed farming systems in Asia. Agricultural Systems. 2002;**71**(1-2):27-40

[42] Martin G, Barth K, Benoit M, Brock C, Destruel M, Dumont B, et al. Potential of multi-species livestock farming to improve the sustainability of livestock farms: A review. Agricultural Systems. 2020;**181**:102821

[43] Cubbage F, Balmelli G, Bussoni A, Noellemeyer E, Pachas AN, Fassola H, et al. Comparing silvopastoral systems and prospects in eight regions of the world. Agroforestry Systems. 2012;**86**:303-314

[44] Broom DM. Components of sustainable animal production and the use of silvopastoral systems. Revista Brasileira de Zootecnia. 2017;**46**:683-688

[45] Aryal DR, Morales-Ruiz DE, López-Cruz S, Tondopó-Marroquín CN, Lara-Nucamendi A, Jiménez-Trujillo JA, et al. Silvopastoral systems and remnant forests enhance carbon storage

in livestock-dominated landscapes in Mexico. Scientific Reports. 2022;**12**(1):16769

[46] Ahmed N, Wahab MA, Thilsted SH. Integrated aquaculture-agriculture systems in Bangladesh: Potential for sustainable livelihoods and nutritional security of the rural poor. Aquaculture Asia. 2007;**12**(1):14

[47] Lazur AM, Leteux F. Integrated aquaculture systems for nutrient reduction in agricultural wastewater: Potential and challenges. Bulletin-Fisheries Research Agency Japan. 2004;**2004**:143-152

[48] Ferreira JML, Maurício RM, Madureira AP. Opportunities and challenges in crop-livestock-forest integration. Informe Agropecuario. 2010;**31**(257):112-120

[49] Dione MM, Traoré I, Kassambara H, Sow AN, Touré CO, Sidibé CAK, et al. Integrated approach to facilitate stakeholder participation in the control of endemic diseases of livestock: The case of Peste des Petits Ruminants in Mali. Frontiers in Veterinary Science. 2019;**6**:392

[50] Behera UK, France J. Integrated farming systems and the livelihood security of small and marginal farmers in India and other developing countries. Advances in Agronomy. 2016;**138**:235-282

[51] Ellis F. Household strategies and rural livelihood diversification. The Journal of Development Studies. 1998;**35**(1):1-38

[52] Shaffril HAM, Krauss SE, Samsuddin SF. A systematic review on Asian's farmers' adaptation practices towards climate change. Science of the Total Environment. 2018;**644**:683-695

[53] Islam MS, Kieu E. Tackling regional climate change impacts and food security issues: A critical analysis across ASEAN, PIF, and SAARC. Sustainability. 2020;**12**(3):883

[54] Raza A, Razzaq A, Mehmood SS, Zou X, Zhang X, Lv Y, et al. Impact of climate change on crops adaptation and strategies to tackle its outcome: A review. Plants. 2019;**8**(2):34

[55] Qaim M. Globalisation of agrifood systems and sustainable nutrition. Proceedings of the Nutrition Society. 2017;**76**(1):12-21

[56] Duchenne-Moutien RA, Neetoo H. Climate change and emerging food safety issues: A review. Journal of Food Protection. 2021;**84**(11):1884-1897

[57] Munang R, Thiaw I, Alverson K, Mumba M, Liu J, Rivington M. Climate change and ecosystem-based adaptation: A new pragmatic approach to buffering climate change impacts. Current Opinion in Environmental Sustainability. 2013;**5**(1):67-71

[58] Godde CM, Mason-D'Croz D, Mayberry DE, Thornton PK, Herrero M. Impacts of climate change on the livestock food supply chain; a review of the evidence. Global Food Security. 2021;**28**:100488

[59] Tefera Y, Hailu Y, Siraj Z. Potential of agroforestry for climate change mitigation through carbon sequestration. Agricultural Research Technology Open Access Journal. 2019;**22**:556196

[60] Mutuo PK, Cadisch G, Albrecht A, Palm CA, Verchot L. Potential of agroforestry for carbon sequestration and mitigation of greenhouse gas emissions from soils in the tropics. Nutrient Cycling in Agroecosystems. 2005;**71**:43-54

[61] Hasimuna OJ, Maulu S, Nawanzi K, Lundu B, Mphande J, Phiri CJ, et al. Integrated agriculture-aquaculture as an alternative to improving small-scale fish production in Zambia. Frontiers in Sustainable Food Systems. 2023;7:1161121

[62] Negash D, Abegaz A, Smith JU, Araya H, Gelana B. Household energy and recycling of nutrients and carbon to the soil in integrated crop-livestock farming systems: A case study in Kumbursa village, Central Highlands of Ethiopia. GCB Bioenergy. 2017;**9**(10):1588-1601

[63] Kumar S, Singh SS, Dey A. Integrated farming systems for eastern India. Indian Journal of Agronomy. 2011;**56**(4):297-304

[64] Phong LT, Udo HM, van Mensvoort ME, Bosma RH, Tri LQ, Nhan DK, et al. Integrated agriculture-aquaculture systems in the Mekong Delta, Vietnam: An analysis of recent trends. Asian Journal of Agriculture and Development. 2007;**4**:51-66

[65] Latawiec AE, Strassburg BB, Valentim JF, Ramos F, Alves-Pinto HN. Intensification of cattle ranching production systems: Socioeconomic and environmental synergies and risks in Brazil. Animal. 2014;**8**(8):1255-1263

[66] Siles P, Cerdán CR, Staver C. Smallholder coffee in the global economy—A framework to explore transformation alternatives of traditional agroforestry for greater economic, ecological, and livelihood viability. Frontiers in Sustainable Food Systems. 2022;**6**:808207

[67] Niroula GS, Thapa GB. Impacts and causes of land fragmentation, and lessons learned from land consolidation in South Asia. Land Use Policy. 2005;**22**(4):358-372

[68] Zhang D, Sial MS, Ahmad N, Filipe AJ, Thu PA, Zia-Ud-Din M, et al. Water scarcity and sustainability in an emerging economy: A management perspective for future. Sustainability. 2020;**13**(1):144

[69] Kristensen E, Jakobsen EB. Challenging the myth of the irrational dairy farmer; understanding decision-making related to herd health. New Zealand Veterinary Journal. 2011;**59**(1):1-7

[70] Stringer LC, Fraser ED, Harris D, Lyon C, Pereira L, Ward CF, et al. Adaptation and development pathways for different types of farmers. Environmental Science & Policy. 2020;**104**:174-189

[71] Hayden J, Rocker S, Phillips H, Heins B, Smith A, Delate K. The importance of social support and communities of practice: Farmer perceptions of the challenges and opportunities of integrated crop–livestock systems on organically managed farms in the northern US. Sustainability. 2018;**10**(12):4606

[72] Hegland SJ. FAO-IPCC Expert Meeting on Climate Change, Land Use and Food Security: Final Meeting Report. 2017

[73] Schut M, Klerkx L, Sartas M, Lamers D, Mc Campbell M, Ogbonna I, et al. Innovation platforms: Experiences with their institutional embedding in agricultural research for development. Experimental Agriculture. United Kingdom. 2016;**52**(4):537-561

[74] Mubvami T, Mushamba S, De Zeeuw H. Integration of agriculture in urban land use planning. Cities farming for the future: Urban agriculture for green and productive cities. In: RUAF, IIRR and IDRC. Silang, the Philippines; 2006. pp. 54-74

[75] Devaux A, Torero M, Donovan J, Horton D. Agricultural innovation and inclusive value-chain development: A review. Journal of Agribusiness in Developing and Emerging Economies. 2018;**8**(1):99-123

[76] Dessart FJ, Barreiro-Hurlé J, Van Bavel R. Behavioural factors affecting the adoption of sustainable farming practices: A policy-oriented review. European Review of Agricultural Economics. 2019;**46**(3):417-471

[77] Ehlers MH, Finger R, El Benni N, Gocht A, Sørensen CAG, Gusset M, et al. Scenarios for European agricultural policymaking in the era of digitalisation. Agricultural Systems. 2022;**196**:103318

[78] Rodriguez JM, Molnar JJ, Fazio RA, Sydnor E, Lowe MJ. Barriers to adoption of sustainable agriculture practices: Change agent perspectives. Renewable Agriculture and Food Systems. 2009;**24**(1):60-71

[79] Mekouar MA. 15. Food and agriculture organization (FAO). Yearbook of International Environmental Law. 2013;**24**(1):587-602

[80] Gebremedhin B, Hoekstra D, Tegegne A. Commercialization of Ethiopian Agriculture: Extension Service from Input Supplier to Knowledge Broker and Facilitator. International Livestock Research Institute; 2006

[81] Sousa F, Nicolay G, Home R. Information technologies as a tool for agricultural extension and farmer-to-farmer exchange: Mobile-phone video use in Mali and Burkina Faso. International Journal of Education and Development using Information and Communication Technology. Burkina Faso. 2016;**12**(3):19-36

[82] Lajoie-O'Malley A, Bronson K, van der Burg S, Klerkx L. The future (s) of digital agriculture and sustainable food systems: An analysis of high-level policy documents. Ecosystem Services. 2020;**45**:101183

[83] Saito N. Mainstreaming climate change adaptation in least developed countries in South and Southeast Asia. Mitigation and Adaptation Strategies for Global Change. 2013;**18**:825-849

[84] Shiferaw B, Hellin J, Muricho G. Improving market access and agricultural productivity growth in Africa: What role for producer organizations and collective action institutions? Food Security. 2011;**3**:475-489

[85] Mushi GE, Di Marzo Serugendo G, Burgi PY. Digital technology and services for sustainable agriculture in Tanzania: A literature review. Sustainability. 2022;**14**(4):2415

[86] Kos D, Kloppenburg S. Digital technologies, hyper-transparency and smallholder farmer inclusion in global value chains. Current Opinion in Environmental Sustainability. 2019;**41**:56-63

Chapter 6

Determination of Grass Quality Using Spectroscopy: Advances and Perspectives

Manuela Ortega Monsalve, Tatiana Rodríguez Monroy, Luis Fernando Galeano-Vasco, Marisol Medina-Sierra and Mario Fernando Ceron-Munoz

Abstract

Spectroscopy is a promising technique for determining nutrients in grasses and may be a valuable tool for future research. This chapter reviews research carried out in recent years, focusing on determining the quality of grasses using spectroscopy techniques, specifically, spectrophotometry. The chemical methods used to determine the nutritional quality of grasses produce chemical residues, are time-consuming, and are costly to use when analyzing large crop extensions. Spectroscopy is a non-destructive technique that can establish the nutritional quality of grass easily and accurately. This chapter aims to describe the techniques focused on the use of spectroscopy and machine learning models to predict and determine the quality of grasses. A bibliographic review was conducted and recent research articles were selected that showed spectroscopic techniques applied to grasses. Different methods and results focusing on the quality of the grasses were compiled. In general, this review showed that the most commonly used spectroscopic method is near-infrared analysis. Spectroscopy is a very effective tool that opens the way to new types of technologies that can be applied to obtain results in determining the quality of pastures, leaving behind the use of traditional methods that represent higher costs and disadvantages compared to traditional methods based on precision agriculture.

Keywords: electromagnetic spectrum, nutrient prediction, precision agriculture, sustainable production, vegetation index

1. Introduction

Grasses and forages are essential for feeding herbivorous animals, especially ruminants that produce milk and meat. In recent years, the growth of the world's population and the resulting need to increase animal food production, have forced grass production to generate a greater amounts of biomass and to seek new technologies that maximize production efficiency. Therefore, it's important to remember that for optimal grass crop production, the soil on which these crops are grown must have good nutrient levels and stable physical characteristics. Therefore, soil characteristics can directly affect the quality and productivity of grass crops. For example, soils that are deficient in one or two nutrients will experience a decrease in grass production [1].

The most important step in determining soil or grass quality is to know the amount of nutrients present. The best way to obtain this information is through a chemical analysis, which indicates nutrient excesses or deficiencies and allows for the development of appropriate fertilization programs [2]. However, the methods used to determine the chemical properties require time, skilled labor, and the use of chemical reagents that are contaminated and dangerous [3]. In addition, according to [4], laboratory analysis presents a number of challenges, such as the analysis of a reduced number of samples and that it can only be performed in the laboratory by destroying the original samples. It also requires a specialized laboratory, a demanding task that is not practical for people working in the field [5]. For this reason, technological alternatives are being considered that are capable of predicting grass properties without generating negative impacts on the environment, and that can be used to control the growth and development of grasses [6]. New calibration techniques and laboratory and remote sensing (RS) belonging to the spectroscopy can predict grasses components without chemical analysis [7].

This chapter aims to describe the techniques focused on the use of spectroscopy and machine learning models that allow the prediction and determination of the quality of grasses. The collection of information on investigative articles was carried out from January 2022 to March 2023 in the databases Google Scholar, ScienceDirect, and SpringerLink. The search terms used were: grasses, soil, spectroscopy, nutrient prediction, agriculture, vegetation index, and machine learning models. Research articles that had a clear methodology and results of using the technique were selected, collecting important information from each of them.

2. Use of spectroscopy in grasses

Spectroscopy captures different portions of the electromagnetic spectrum (ES). The light variations in the ES regions cover the visible spectrum (VIS) consisting of bands like red (R), green (G), and blue (B) and the infrared (IR) region which is divided into the near-infrared (NIR), short wave infrared (SWIR), medium wave infrared (MWIR) and long wave infrared (LWIR) spectra as shown in **Figure 1**. The VIS and IR region of the ES are the most used equipment that detects pasture quality by spectroscopy.

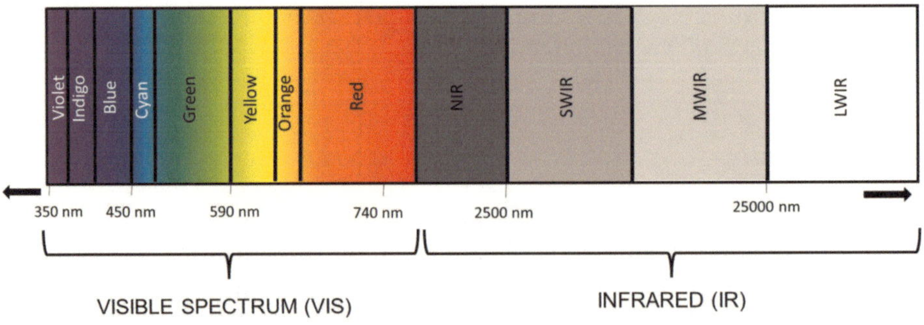

Figure 1.
Visible and infrared regions of the electromagnetic spectrum. Adapted to [8].

Among the spectroscopic techniques is spectrophotometry, a branch based on the use of different parts of the electromagnetic spectrum and the presence of light to detect different elements according to their chemical composition. Among the methods using spectrophotometry we can find cameras with NIR, RGB, multispectral (MSI), hyperspectral (HSI) and RS bands. NIR works with vibrations [9, 10]; and according to [11], this technique has gained attention because it's widely used in different areas of science, proving to be successful in obtaining information that is difficult to obtain with competing techniques. Now, spectroscopy uses the MWIR zone, which is known as a vibrational method that uses the light emitted in a sample to measure the transmission, absorption and reflection of light. This technique detects the structures of the sample as the molecules have chemical bonds that generate vibrational energy [12, 13]. In the same way, HSI detects signals or contiguous bands that contain a very narrow spectral bandwidth [14], greatly facilitate the detection of sample elements [15]. However, the use of HSI has not been sufficiently explored due to it's high acquisition cost and the large amount of spectral data that must be processed [16, 17]. The difference between these methods is the range of the spectrum they use to determine different types of features related to the spectral bands.

Cameras that record the spectrum's bands can be mounted on unmanned aerial vehicles (UAVs) or placed permanently in a laboratory to record data from samples. UAVs are widely used to collect data in the field due to their simplicity of operation and on-site data collection. For this reason, [7] used UAVs with HSI cameras to evaluate quality characteristics in grasses and demonstrated that this method can predict forage quality parameters.

Figure 2 shows two ways to analyze the quality of grasses through different forms that use spectrophotometry.

The components of a sample can be analyzed from great distances using satellites; this is known as RS. RS has been used to study the spectral, spatial, and temporal variations of electromagnetic waves, and revealed the correlations between them and the proprieties of different terrestrial materials [18]. In other words, it focuses on identifying the materials present on the earth's surface and on understanding the

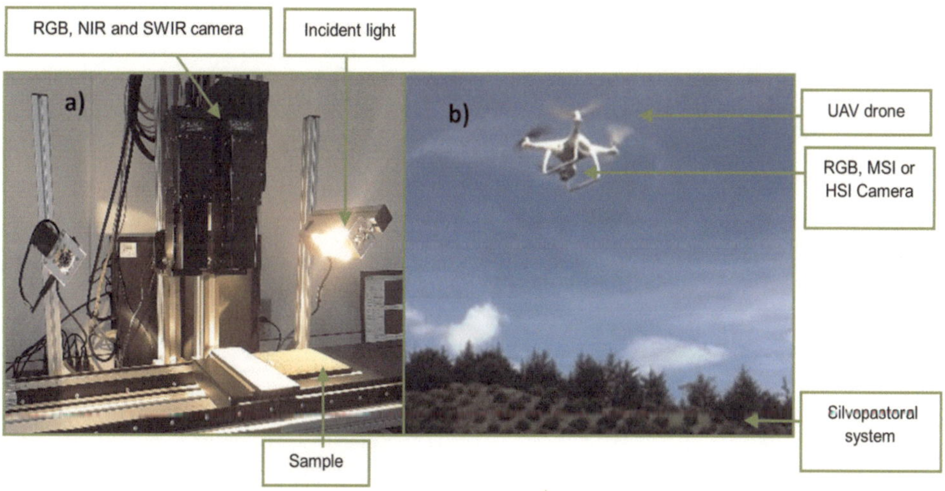

Figure 2.
Different forms to apply spectroscopy to determine the quality of grasses: (a) hyperspectral images in the laboratory and (b) UAV drone with RGB, multispectral or hyperspectral moving cameras.

phenomena that occur on it through their spectral signature, that is, it seeks to establish a relationship between the properties of the light reflected or emitted by terrestrial objects and the intrinsic properties of these objects [19]. To perform this type of data analysis, RS analysis and processing techniques are used, using satellites, to capture electromagnetic radiation at different wavelengths [20, 21]. All of this provides valuable information about the composition and dynamics of the land surface allowing producers to detect crop variations and facilitate a more precise application of fertilizers [22] nutritional composition and general condition of plants.

It's important to note that plants exhibit dynamic behavior in ES due to phenological changes in the plant, lighting conditions associated with topography (slope and orientation), sun position over the year, and soil moisture conditions, which can result in significant variations in the spectral response pattern [23].

The spectral signature of plants is characterized by how they reflect light in the visible spectrum (400–700 nm). Chlorophyll absorbs strongly in the visible spectrum around the B and R bands, resulting in a low reflectance, and the G spectral band, the reflectance is higher. It's in this region of the spectrum that photosynthesis occurs, a process in which energy is absorbed [24]. On the other hand, in the NIR region, the vegetation shows its highest reflectance in the NIR between 700 and 1300 nm, around 45–50% (**Figure 3**). This phenomenon is the result of diffusion caused by the refractive indices of the intracellular fluid and the intercellular spaces present in the plant mesophyll, especially in the spongy mesophyll. In contrast, at wavelengths between 1300 and 2500 nm, the reflectivity of the sheet it's controlled by water absorption, resulting in reflectance values of 10–20% [18]. When the vegetation is mature or under stress due to disease, insect attack, or low humidity, changes in the spectral characteristics of the leaves occur [25]. In general, these changes occur simultaneously in the VIS and IR regions, but there are greater alterations in IR. This behavior explains the great utility of ES for the study of vegetation.

Figure 3 shows the spectral signature of the vegetation under different phenological conditions. It can be observed that the reflectance is influenced by the

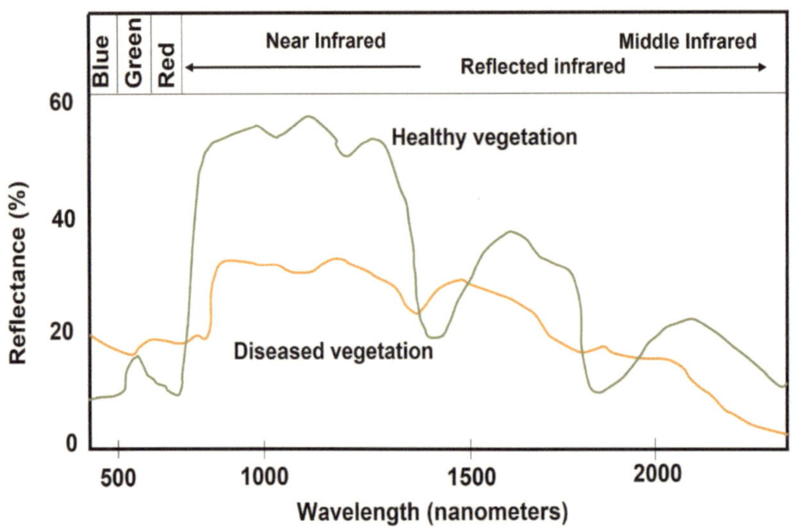

Figure 3.
Changes in the spectral signature of healthy and diseased vegetation in the visible and infrared portions of the electromagnetic spectrum. Adapted to [19].

concentration of pigments in the leaves, mainly chlorophyll and carotenoids. In the VIS region of the spectrum, reflectance and transmittance are low due to strong absorption by leaf pigments.

2.1 Vegetation indices in grasses

From the study of spectroscopy, numerous vegetation indices (VI) have been developed. These indices are defined as parameters calculated using reflectance values at different wavelengths, to extract information related to vegetation [26]. In this context, VI is generated by mathematical calculations involving the different spectral bands of the images. These calculations produce a new image that highlights specific characteristics related to the physiological functioning of the plants [27]. The VI defined so far have in common the use of reflectance values in the R and NIR spectral bands, which are dedicated to the spectral behavior in this region of the ES. The reflectance of the vegetation goes from a relative minimum in the R corresponding to the absorption band of chlorophyll to an absolute maximum in the NIR, which is a consequence of the multiple scattering of the radiation within the cellular structure [25].

One of the most commonly used indices in vegetation assessment is the normalized difference vegetation index (NDVI), which quantifies the relationship between the energy absorbed and emitted by terrestrial objects [28]. In addition to NDVI, other indices relate to biomass and leaf area index. These include the green normalized difference vegetation index (GNDVI), which uses the G band instead of the R-band of the spectrum. There is also the RATIO index, which relates the high NIR reflectance of vegetation to low R reflectance. The enhanced vegetation index (EVI) is an enhancement of NDVI designed to perform better in areas of dense vegetation and to reduce atmospheric effects by being sensitive to variations in vegetation cover and leaf area. Similarly, the adjusted ground vegetation index (SAVI) is an enhancement of NDVI that attempts to compensate for the effects of ground brightness by relating the nutrients and R bands [29, 30]. Another index used in grasses is the normalized pigmented chlorophyll ratio index (NPCI) [31]. The VI is useful in determining productive characteristics in grasses, such as height, presence of disease, or available quantity.

Table 1 shows the most commonly used vegetation indices used to evaluate biomass production and nutritional content of grasses, and the results obtained when applying these indices. They are mainly used to determine the productive properties that can directly influence the quality of the plants and their use in animal feed.

Among the results found to evaluate the quality and nutritional content of forages by spectroscopy, the work of [40] stands out. The reason is that this study has found a significant correlation between forage yield and RGB bands based on VI. This positive relationship demonstrates the usefulness of VI derived from color imagery for estimating forage quality and nutrient content. Soil chemical characteristics were also evaluated as an indirect measure of grass quality. One study found that soil nutrient content, especially potassium and phosphorus, was strongly related to electrical conductivity and NDVI [41]. These results demonstrate the relationships between soil and crop characteristics and allow the optimization of variable rate fertilization.

Likewise, Fava et al. [33] evaluated biomass and nitrogen status in grasses at three different growth stages and in grazed and ungrazed plots, finding good results for assessing nitrogen content, green biomass, and leaf area index using the NIR (775–820 nm) and longer wavelengths of the red edge (740–770 nm). Another report shows the feasibility of using NDVI to assess moisture content in forages.

Vegetation index	Formula	Characteristics evaluated	Author
NDVI	NIR – R/NIR + R	Type of grass and amount	[32]
RATIO	NIR/R	Biomass and nitrogen	[33]
GNDVI	NIR – G/NIR + G	Differentiation between soil and grasses	[34]
NDRE	NIR – RE/NIR + RE	Amount and chlorophyll	[35]
SAVI	NIR – R/NIR + R + 0.5	Amount and cover	[36, 37]
OSAVI	1.6 * (NIR – R)/(NIR + R + 0.16)	Biomass	[38]
EVI	2.5 * (NIR–R)/(NIR + 6 * R – 7.5 * B) + 1	Biomass and cover	[39]

Vegetation index: NDVI, normalized difference; RATIO, simple ratio indices; GNDVI, green normalized difference; NDRE, normalized difference red edge; SAVI, soil adjusted; OSAVI, optimized soil adjusted; EVI, enhanced; R, red; G, green; B, blue; NIR, near infrared; RE, red edge.

Table 1.
Vegetation index, formulas, constants and characteristics were evaluated to predict the quality of pastures through spectroscopic techniques.

2.2 Quality characteristics in grasses determined through spectroscopic techniques

In general, the quality of grasses is measured using various characteristics such as dry matter, digestibility, energy, organic matter, protein, and carbohydrates [42, 43]. Also, by many VI, factors affecting the quality of grasses, such as plant diseases, amount of biomass, or type of fertilization, are directly related to the nutritional characteristics (**Figure 4**). In addition, the quality of the grasses can also be measured indirectly, through the content of soil nutrients. In the soil matrix, the content of sand, silt, and clay can be monitored by spectroscopy [44, 45], the content of organic carbon and organic matter, which are synonymous with soil fertility and therefore, of plant growth [46, 47]. However, several studies have managed to determine using spectroscopy many characteristics focused on the productivity of grasses.

Results of previous studies agree that the use of spectroscopy in forages can be useful to determine factors such as mass and growth, light interception, turf heterogeneity, nitrogen deficiency, and drought stress, using non-destructive methods [48]. The determination of the concentration of nutrients such as phosphorus, potassium,

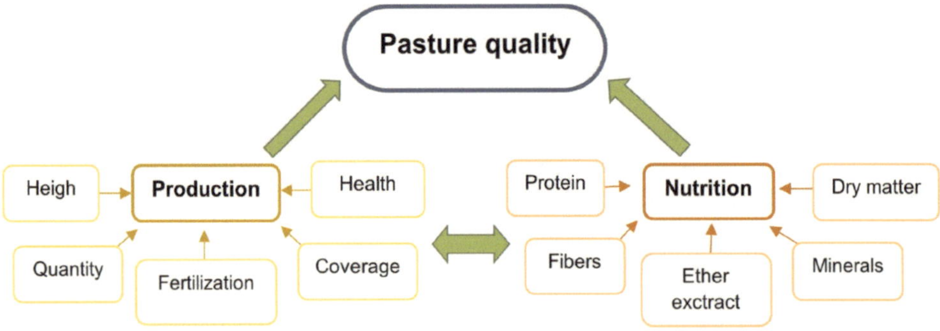

Figure 4.
Main nutritional and production quality characteristics evaluated in grasses.

sulfur, calcium, and magnesium, some minor elements. and the sugar content of grasses has also been studied using spectroscopic techniques [49]. In addition, spectroscopy can work in other areas indirectly related to the quality of grasses and plants, for example, [50] they worked on forest dynamics and land use; in addition [51] applied spectroscopic techniques to determine the quality of the milk from cattle fed with fresh grasses and [52] evaluated carcasses of lambs that were fed with fresh grasses. This is due to the fact that there are many characteristics, both direct and indirect, that can be evaluated in grasses to determine their quality. This can open new avenues for future research and can clarify that these techniques, belonging to the new PA, are widely used in the investigative field of vegetation.

2.3 Investigations carried out in grasses using spectroscopic techniques

A study conducted by [53] demonstrates the implementation of portable spectroscopic sensors to determine the nutrient content in grasses, finding good results in determining height and protein content, which can lead to improved productivity for farmers through access to this type of technology. In this review, it has been shown that NIR instruments are the most widely used in spectroscopy and many of the studies carried out in grasses focus on the use of this method, with good results. For example, Restaino et al. [54] used 120 grass samples analyzed by NIR to determine nutrients in grasses, demonstrating great potential in this technique; Danelli et al. [55] predicted grass quality parameters using NIR, using 1615 samples, managed to find optimal settings, and considered this technique as an aid in grass management; Catunda et al. [56] developed calibration strategies in grasses using NIR to predict their nutritional composition using 2622 samples, they found that characteristics such as ash, protein, and fiber can be easily determined using this methodology; Parrini et al. [57] evaluated the quality of fresh grasses using NIR, finding the estimation of some chemical parameters of grasses associated with quality feasible.

The NIR methodology has the advantage that it can be applied in the laboratory and in the field, for this reason, Serrano et al. [5] used grasses for continuous grazing to test the technique in the field, demonstrating that the combination of spectroscopic and ultrasonic methods to determine pasture quality factors can be accurate even in very heterogeneous grasses due to grazing animal. In this review, 95% of the studies carried out have included NIR analysis in the laboratory or through UAVs, where methodologies that include the use of HSI and multispectral images can also be observed, while the other percentage have used techniques that include only RGB bands. This leads to the idea that studies of grasses have evolved to include not only the visible part of the spectrum but also using the NIR region, which provides more information due to its greater number of spectral bands.

Table 2 summarizes the information found by different authors on the use of spectroscopy in grasses to determine their quality, including parameters such as the method used, the spectral characteristics of the method, and the results obtained in this review.

2.4 Machine learning models and predictive metrics applied to grass quality

In the area of measuring the intensities of reflected light in narrow spectral bands, particularly in the NIR range, advances in optical methods have enabled the identification of chemical bonds between hydrogen and carbon, hydrogen and nitrogen, and hydrogen and oxygen. These chemical bonds have proven to be key indicators for

Study	Spectral range	Results
Techniques in real time and monitor the changes in the nutrients of the different types of grasses with NIR portable [53]	950–1750 nm	High correlation between grass height and some nutrients
Potential data for grass quality assessment using UAV with HSI in field [7]	350–2500 nm	Models with good predictive potential for CP and fibers
Spectroscopy for predicting the nutritional value of forage silages with NIR [54]	400–2500 nm	Potential to predict DM, CP and fibers
Calibrations for predicting fresh grass quality parameters with NIR [55]	1100–2500 nm	Optimal predictions for fibers and ash variables
The best calibrations for a grassland sample collection with NIR [5]	1100–2500 nm	Adjustments greater than 80% for pasture nutritional variables
Predictive models for estimating the chemical composition of Brachiaria with NIR [58]	400–2500 nm	Nutrient variables were estimated with higher prediction in dry grasses
Use canopy spectra with NIR portable to predict parameters for individual ryegrass plants [59]	300–2500 nm	It was possible to perform a classification of the plants using of VI
Predictive models to estimate different concentrations at the field level, using spectroscopy data from grass images using UAV with HSI [60]	450–998 nm	Generation of predictive models for the CP and fibers variables and their variations in space
Performance of DL-based segmentation algorithms in the context of crop segmentation approaches using UAV with RGB [61]		It was possible to determine different compositions of plants using models
Design and development of a portable using UAV with MSI for grassland vegetation growth monitoring system [6]	450, 550, 650, 750, 850, and 960 nm	The growth of different pastures was predicted using VI

SR, spectral range; HSI, hyperspectral; CP, crude protein; DM, dry matter; DL, deep learning; MSI, multispectral; UAV, unmanned aerial vehicles.

Table 2.
Compilation of different spectral methodologies and characteristics of the equipment used in spectroscopy applied to grasses.

determining the nutritional characteristics of grass, including the presence of crude protein, fibers, and other plant constituents [57, 59, 62]. These new technological alternatives, along with the use of machine learning statistical models, can produce good results in predicting various nutrient variables in soils and grasses. According to [63], machine learning algorithms have great potential for analyzing spectral data and analyzing attributes in grasses. Statistical models perform the function of developing, evaluating, and improving the equations that calibrate and predict the presence of elements or nutrients from reflectance spectra, replacing traditional chemical techniques [64].

Knowing the relationship between the spectral bands and the grass variables requires the use of multivariate statistics [65], so several statistical models have been developed. The most commonly used is the partial least squares regression (PLSR), which explores the relationships that may exist between all the variables [66]. The cubist model (CUB) is also used to predict nutrients using spectroscopy. This algorithm performs the construction of an unconventional regression tree [67], the prediction is made based on intermediate linear models step by step, and it creates subsets of data and rules to select only some predictor variables [68]. On the other hand, the random forest (RF) model works with many decision trees that are known as classifiers; these use representative variables of the sample, forming classifier nodes for the set of variables that have multivariate characteristics, in the case of the spectral bands of each sample [69].

In addition, the support vector machine (SVM) algorithm, using the hyperplane, recognizing different categories, and maximum margins, manages to separate the data into different categories by selecting the most appropriate vectors for the prediction [70]. However, there are many algorithms that can be applied to spectroscopic data. Others widely used are principal component regression (PCA), multiple linear regression (MLR), and artificial neural networks (ANN) [71, 72]. To evaluate the predictive performance of the above machine learning models, [73] considers that some of the most commonly used metrics in these models are the coefficient of determination (R^2), the ratio of performance to deviation (RPD), and the root-mean-square error (RMSEP). It's important to obtain models that have R^2 values close to 1, low RMSEP values since the difference between the predicted and observed values of the model should be small, and RPD values greater than 2 [74]. These model selection parameters will indicate that the selected model can efficiently predict grass quality properties using spectroscopic methods.

In reviewing the selected articles, it was found that 84% of the articles belonging to this review used the PLSR machine learning model for data analysis. The remaining percentage includes PCA, SVM, and CUB models. The widespread use of the PLSR model for predicting quality variables using spectroscopy may be due to its ease of execution in various statistical software, in addition to the fact that it is a model that reduces the dimensionality of the data, and can reduce memory requirements.

For this reason, the PLSR models used in the various studies collected in this review presented optimal fits in predicting different characteristics in the grasses, providing high reliability for use in future studies. For example, [75] used this model to predict crude protein (CP) in grasses and found an R^2 of 0.98 and an RPD of 4.12, indicating a high fit of the model found. Results of a R^2 were also found to predict the CP in the grass of 0.69 and 0.73 respectively, but the RPD found was 1.95, which means that the performance of the model for this feature is regular [76, 77]. Another similar result was found by [78], where the RPD found was 1.84, the R^2 was 0.77 and the RMSEP was 2.05. Finally, for the CP variable, [5] found an R^2 of 0.84 and an RPD of 4. According to all the results related to this nutritional variable of the grass, the CP can be predicted with PLSR models with a high fit, but many samples must be evaluated to make the model more and more fit. The fibers variable was also been predicted using PLSR models, [75] finding an R^2 of 0.94, an error of 2.94, and an RPD of 2.83. For the same variable, R^2 of 0.67 and 0.76 and RMSEP of 5.77 and 4 were obtained [7, 56]. This variable was predicted using RS and found a fit of 0.66 and an RPD of 1.71 [79]. This variable resulted in minor adjustments to the CP variable but has great potential to be predicted using spectroscopy and machine learning models.

Regarding other variables that have been predicted in grasses, [55] managed to find an R^2 for the ash characteristic in the grass of 0.75 and an error of 1.01, which means that this technique allows one to know the amount of minerals in the grass. Also, [80] found a good fit for the nutrient variable EE in the grass, finding an R^2 of 0.73, an RDP of 1.69, and an RMSEP of 0.23. The CP, fibers, and dry matter variables are the most analyzed by researchers and the CP variable is the one that shows the highest adjustment and prediction results in the different evaluation criteria of the models. However, the other nutritional variables also present good adjustment results.

Regarding the productive variables of grasses, VI has also been widely used in spectroscopy to determine nutritional and productive variables, mainly in remote sensing and moving cameras. In general, the variables related to the quality of the pasture, evaluated using spectroscopy and statistical analysis, have generated optimal results in terms of their adjustment, which indicates that the use of these techniques is a great tool for knowing the quality of the grass, before being offered as food to cattle. Regarding the statistical models, although SVM and CUB have been used in this type

Parameter	Statistical model	R^2	RMSE	RPD	Author
Crude protein	PLSR	0.79	2.48	—	[7]
		0.90	—	2.37	[81]
		0.88	2.14	2.26	[57]
		0.84	1.62	4.00	[5]
		0.90	1.48	3.1	[82]
		0.83	0.64	2.4	[80]
		0.77	2.05	1.84	[78]
	PCA	0.82	—	1.88	[58]
	RF	0.70	2.95	—	[7]
Dry matter	PLSR	0.96	—	2.60	[81]
		0.87	2.70	2.75	[57]
		0.94	1.97	2.7	[82]
		0.57	—	1.51	[58]
Fibers	PLSR	0.60	6.34	—	[7]
		0.76	4.00	1.54	[57]
		0.89	2.47	2.90	[82]
		0.30	3.32	1.1	[82]
		0.55	3.23	1.50	[80]
		0.61	6.98	1.90	[5]
	RF	0.52	6.98	—	[7]
	PCA	0.82	—	2.23	[61]
Ash	PLSR	0.64	1.05	1.87	[58]
		0.75	1.01	1.2	[82]
		0.71	0.63	1.87	[80]
Ether extract	PLSR	0.45	0.28	1.18	[57]
		0.65	0.24	1.4	[82]
		0.73	0.23	1.69	[80]
Lignine	PLSR	0.73	1.83	1.61	[57]

PLSR, partial least squares; PCA, principal component analysis; RF, random rorest; R^2*, determination coefficient; RMSEP, root-mean-square error; RPD, ratio of performance to deviation.*

Table 3.
Nutritional characteristics evaluated in grasses using spectroscopy.

of study, they are not preferred by researchers, and it is then, the PLSR model is the most used. The NIR methodology is the most used to carry out research on grass quality due to its wide range in the electromagnetic spectrum, NIR the interaction between plant compounds can be observed much easier, while equipment that only RGB cameras or multispectral may be limited in predicting plant properties due to few bands and little information available. This is due to the great simplicity of programming and interpretation of the results, as it reduces the dimensionality of the spectral data. **Table 3** shows a summary of the studies carried out on the grasses, the statistical models applied and the fitting values found in each one of them.

3. Conclusions

Spectroscopy has been widely used in recent years to evaluate the quality of grasses. This is confirmed by the large amount of research supporting its use and demonstrating its effectiveness in predicting productive and nutritional variables in grasses. Both laboratory and field methods can be easily applied, and the use of machine learning statistical models to predict variables related to quality has promoted its use in agricultural areas. In line with the rapid advances that have occurred in the field of precision agriculture, it is expected that the development and use of spectroscopic techniques to determine grass quality will increase in the coming years due to the need to limit the use of chemical analysis and promote non-destructive methods in the field. The development and widespread use of more accessible machine learning statistical models and portable equipment is expected to enable the collection and analysis of real-time information in the field to facilitate the work of graziers.

Acknowledgements

This publication and the financial support of the MSc students was possible thanks to the project "Design and validation of predictive models to determine Cation Exchange Capacity (CEC), Organic Matter (OM) and Nitrogen (N) in soils from hyperspectral images" through the agreement 2022-7204, financed by the University of Antioquia Foundation.

Conflict of interest

The authors declare no conflict of interest.

List of abbreviations

NIR	near infrared
RS	remote sensing
ES	electromagnetic spectrum
VIS	visible spectrum
IR	infrared
R	red

G	green
B	blue
RE	red edge
SWIR	short wave infrared
MWIR	medium wave infrared
LWIR	long wave infrared
UAVs	unmanned aerial vehicles
HSI	hyperspectral
MSI	multispectral
VI	vegetation index
NDVI	normalized difference vegetation index
GNDVI	green normalized difference vegetation index
RATIO	simple ratio index
NDRE	normalized difference red edge
EVI	enhanced vegetation index
SAVI	adjusted ground vegetation index
NPCI	chlorophyll ratio index
PLSR	partial least square regression
RF	random forest
SVM	support vector machine
CUB	cubist model
R^2	coefficient of determination
RMSEP	root-mean-square error
RPD	ratio of performance to deviation
CP	crude protein
DM	dry matter

Author details

Manuela Ortega Monsalve[1*], Tatiana Rodríguez Monroy[1], Luis Fernando Galeano-Vasco[2], Marisol Medina-Sierra[2] and Mario Fernando Ceron-Munoz[2]

1 Animal Sciences, GAMMA Research Group, Faculty of Agricultural Sciences, Antioquia University, Medellin, Antioquia, Colombia

2 GAMMA Research Group, Faculty of Agricultural Sciences, Antioquia University, Medellin, Antioquia, Colombia

*Address all correspondence to: manuela.ortegam@udea.edu.co

References

[1] Silveira ML, Kohmann MM. Maintaining soil fertility and health for sustainable pastures. In: Management Strategies for Sustainable Cattle Production in Southern Pastures. Cambridge, Massachusetts: Academic Press; 2020. pp. 35-58. DOI: 10.1016/b978-0-12-814474-9.00003-7

[2] Westerman RL. Soil Testing and Plant Analysis. 3rd ed. Vol. 152, No. 2. Madison: Soil Science; 1991. p. 137

[3] Molano ML. Caracterización nutricional de forrajes tropicales usando espectroscopía de infrarrojo cercano (nirs). Bogotá, Colombia: Universidad Nacional de Colombia; 2013

[4] Teye E, Amuah CL, Atiah K, Darko RO, Abindaw T, Amoah KK, et al. Quick determination of soil quality using portable spectroscopy and efficient multivariate techniques. Journal of Spectroscopy. 2023;**2023**:11. DOI: 10.1155/2023/2024318

[5] Serrano J, Shahidian S, Carreira E, Nogales-Bueno J, Rato AE. Predicting the evolution of pasture quality by NIRS: Perspectives for real-time pasture and grazing management. In: Online AgEng2021 Conference, 5–8 July 2021. Évora, Portugal; 2021. pp. 1-8

[6] Zhang A, Hu S, Zhang X, Zhang T, Li M, Tao H, et al. Ahandheld grassland vegetation monitoring system based on multispectral imaging. Agriculture. 2021; **11**(12):12-62. DOI: 10.3390/agriculture11121262

[7] Fernández Habas J, Cañada MC, Moreno AM, Leal-Murillo JR, González-Dugo MP, Oar BA, et al. Estimating pasture quality of Mediterranean grasslands using hyperspectral narrow bands from field spectroscopy by random forest and PLS regressions. Computers and Electronics in Agriculture. 2022;**192**:106614. DOI: 10.1016/j.compag.2021.106614

[8] Akhloufi M, Bendada A. Fusion of active and passive infrared images for face recognition. Proceedings of SPIE - The International Society for Optical Engineering. 2013;**8705**:84-93. DOI: 10.1117/12.2017942

[9] Huck CW, Bec KB. Advances in near-infrared spectroscopy and related computational methods. Molecules. 2019;**24**(23):43-70. DOI: 10.3390/books978-3-03928-053-7

[10] Czarnecki MA, Morisawa Y, Futami Y, Ozaki Y. Advances in molecular structure and interaction studies using near-infrared spectroscopy. Chemical Reviews. 2015;**115**(18): 9707-9744. DOI: 10.1021/cr500013u

[11] Bec KB, Grabska J, Huck CW. Near-infrared spectroscopy in bio-applications. Molecules. 2020;**25**(12): 29-48. DOI: 10.3390/molecules25122948

[12] Coates J. Interpretation of infrared spectra, a practical approach. In: Meyers RA, McKelvy ML, editors. Encyclopedia of Analytical Chemistry: Applications, Theory and Instrumentation. Chichester, England; 2000. DOI: 10.1002/9780470027318.a5606

[13] Amirvaresi A, Nikounezhad N, Amirahmadi M, Daraei B, Parastar H. Comparison of near-infrared (NIR) and mid-infrared (MIR) spectroscopy based on chemometrics for saffron authentication and adulteration detection. Food Chemistry. 2021;**344**: 128647. DOI: 10.1016/j.foodchem.2020.128647

[14] Randelović P, Dorević V, Milić S, Balesević-Tubić S, Petrović K, Miladinović J, et al. Prediction of soybean plant density using a machine learning model and vegetation indices extracted from RGB images taken with a UAV. Agronomy. 2020;**10**(8):1108. DOI: 10.3390/agronomy10081108

[15] Lu B, Dao PD, Liu J, He Y, Shang J. Recent advances of hyperspectral imaging technology and applications in agriculture. Remote Sensing. 2020; **12**(16):26-59. DOI: 10.3390/rs12162659

[16] Darvishzadeh R, Matkan AA, Ahangar AD. Inversion of a radiative transfer model for estimation of rice canopy chlorophyll content using a lookup-table approach. Journal of Selected Topics in Applied Earth Observations and Remote Sensing. 2012; **5**(4):1222-1230. DOI: 10.1109/JSTARS.2012.2186118

[17] Lodhi V, Chakravarty D, Mitra P. Hyperspectral imaging system: Development aspects and recent trends. Sensing and Imaging. 2019;**20**(35):1-24. DOI: 10.1007/s11220-019-0257-8

[18] Chuvieco E. Fundamentos de teledetección espacial. Estudios Geográficos. 1991;**52**(203):371

[19] Mena FC. Tecnologías geoespaciales: experiencias aplicadas al estudio y gestión del territorio. 2009

[20] Paoletti ME, Haut JM, Plaza J, Plaza A. Estudio comparativo de técnicas de clasificación de imágenes hiperespectrales. Revista Iberoamericana de Automática e Informática Industrial. 2019;**16**(2):129-137. DOI: 10.4995/riai.2019.11078

[21] Mantripragada K, Dao PD, He Y, Qureshi FZ. The effects of spectral dimensionality reduction on hyperspectral pixel classification: A case study. PLoS One. 2022;**17**(7):0269174. DOI: 10.1371/journal.pone.0269174

[22] Su WH. Advanced machine learning in point spectroscopy, RGB-and hyperspectral-imaging for automatic discriminations of crops and weeds: A review. Smart Cities. 2020;**3**(3):767-792. DOI: 10.3390/smartcities3030039

[23] Houborg R, Fisher JB, Skidmore AK. Advances in remote sensing of vegetation function and traits. International Journal of Applied Earth Observation and Geoinformation. 2015; **43**:1-6. DOI: 10.1016/j.jag.2015.06.001

[24] Gausman HW. The leaf mesophylls of twenty crops, their light spectra, and optical and geometrical parameters. 1972. 10.22004/ag.econ.160035.

[25] Kharuf-Gutierrez S, Hernández S, Orozco R, Aday OC, Delgado I. Análisis de imágenes multiespectrales adquiridas con vehículos aéreos no tripulados. Ingeniería Electrónica, Automática y Comunicaciones. 2018;**39**(2):79-91

[26] Tucker CJ. Red and photographic infrared linear combinations for monitoring vegetation. Remote Sensing of Environment. 1979;**8**(2):127-150. DOI: 10.1016/0034-4257(79)90013-0

[27] Gilabert MA, González-Piqueras J, García-Haro J. Acerca de los índices de vegetación. Revista de Teledetección. 1997;**8**:1-10

[28] Pettorelli N. The Normalized Difference Vegetation Index. USA: Oxford University Press; 2014. p. 224. DOI: 10.1093/acprof:osobl/9780199693160.001.0001

[29] Bannari A, Morin D, Bonn F, Huete AR. A review of vegetation indices.

Remote Sensing Reviews. 1995;**13**(1–2): 95-120. DOI: 10.1080/02757259509532298

[30] Rondeaux G, Michael S, Baret F. Optimization of soil-adjusted vegetation indices. Remote Sensing of Environment. 1996;**55**(2):95-107. DOI: 10.1016/0034-4257(95)00186-7

[31] Tao H, Feng H, Xu L, Miao M, Largo H, Yue J, et al. Estimation of crop growth parameters using UAV-based hyperspectral remote sensing data. Sensors. 2020;**20**(5):1296. DOI: 10.3390/s20051296

[32] Rouse J, Haas R, Schell J, Deering D. Monitoring vegetation systems in the Great Plains with ERTS. NASA Special Publication. 1974;**351**(1):309-317

[33] Fava F, Colombo R, Bocchi S, Meroni M, Sitzia M, Fois N, et al. Identification of hyperspectral vegetation indices for Mediterranean pasture characterization. International Journal of Applied Earth Observation and Geoinformation. 2009;**11**(4): 233-243. DOI: 10.1016/j.jag.2009.02.003

[34] Huete AR. A soil-adjusted vegetation index (SAVI). Remote Sensing of Environment. 1988;**25**(3):295-309

[35] Barnes E, Clarke T, Richards S, Colaizzi P, Haberland J, Kostrzewski M, et al. Coincident detection of crop water stress, nitrogen status and canopy density using ground based multispectral data. In: Proceedings of the Fifth International Conference on Precision Agriculture; Bloomington, USA. 2000

[36] Daughtry C, Walthall C, Kim M, De Colstoun EB, McMurtrey IJ. Estimating corn leaf chlorophyll concentration from leaf and canopy reflectance. Remote Sensing of Environment. 2000;**74**(2): 229-239. DOI: 10.1016/S0034-4257(00)00113-9

[37] Qi J, Chehbouni A, Huete AR, Kerr YH, Sorooshian S. A modified soil adjusted vegetation index. Remote Sensing of Environment. 1994;**48**(2): 119-126. DOI: 10.1016/0034-4257(94)90134-1

[38] Boschetti M, Bocchi S, Brivio PA. Assessment of pasture production in the Italian Alps using spectrometric and remote sensing information. Agriculture, Ecosystems and Environment. 2007;**118** (1–4):267-272. DOI: 10.1016/j.agee.2006.05.024

[39] Jiang Z, Huete AR, Didan K, Miura T. Development of a two-band enhanced vegetation index without a blue band. Remote Sensing of Environment. 2008;**112**(10):3833-3845. DOI: 10.1016/j.rse.2008.06.006

[40] Lussem U, Bolten A, Gnyp ML, Jasper J, Bareth G. Evaluation of rgb-based vegetation indices from UAV imagery to estimate forage yield in grassland. Developments, Technologies and Applications in Remote Sensing. 2018;**42**:1215-1219. DOI: 10.5194/isprs-archives-XLII-3-1215-2018

[41] Mazur P, Gozdowski D, Wójcik-Gront E. Electrical conductivity of the soil and vegetation indices derived from satellites for the evaluation of the content of phosphorus, potassium and magnesium, pH and delimitation of management zones within the field. Agriculture. 2022;**12**(2):354. DOI: 10.3390/agricultura12060883

[42] O'Donovan M, Lewis E, O' Kiely P. Requirements of future grass-based ruminant production systems in Ireland. Irish Journal of Agricultural and Food Research. 2011;**50**(1):1-21

[43] Burns GA, Gilliland TJ, Grogan D, Watson S, O' Kiely P. Assessment of herbage yield and quality traits of

perennial ryegrasses from a national variety evaluation scheme. The Journal of Agricultural Science. 2013;**151**(3):331-346. DOI: 10.1017/S0021859612000251

[44] Jaconi A, Cora V, Axel D. Near-infrared spectroscopy as an easy and precise method to estimate soil texture. Geoderma. 2019;**337**:906-913. DOI: 10.1016/j.geoderma.2018.10.038

[45] Coblinski JA, Giasson É, Demattê JA, Dotto AC, Costa JJF, Vašát R. Prediction of soil texture classes through different wavelength regions of reflectance spectroscopy at various soil depths. Catena. 2020;**189**:104485. DOI: 10.1016/j.catena.2020.104485

[46] Crucil G, Castaldi F, Aldana-Jague E, van Wesemael B, Macdonald A, Van Oost K. Assessing the performance of UAS-compatible multispectral and hyperspectral sensors for soil organic carbon prediction. Sustainability. 2019; **11**(7):1889. DOI: 10.3390/su11071889

[47] Meng X, Bao Y, Liu J, Liu H, Zhang X, Zhang Y, et al. Regional soil organic carbon prediction model based on a discrete wavelet analysis of hyperspectral satellite data. International Journal of Applied Earth Observation and Geoinformation. 2020;**89**(1): 102-111. DOI: 10.1016/j.jag.2020.102111

[48] Zhou W, Yang H, Xie L, Li H, Huang L, Zhao Y, et al. Hyperspectral inversion of soil heavy metals in three-river source region based on random forest model. Catena. 2021;**202**:105222. DOI: 10.1016/j.catena.2021.105222

[49] Schut AGT, Ketelaars JJMH. Monitoring growth of grass swards using imaging spectroscopy. Grass and Forage Science. 2003;**58**(3):276-286. DOI: 10.1046/j.1365-2494.2003.00379.x

[50] Schut AGT, Lokhorst C, Hendriks MMWB, Kornet JG, Kasper G. Potential of imaging spectroscopy as tool for pasture management. Grass and Forage Science. 2005;**60**(1):34-45. DOI: 10.1111/j.1365-2494.2005.00449.x

[51] Herpin U, Cerri CC, Carvalho MCS, Markert B, Enzweiler J, Friese K, et al. Biogeochemical dynamics following land use change from forest to pasture in a humid tropical area (Rondonia, Brazil): A multi-element approach by means of XRF-spectroscopy. Science of the Total Environment. 2002;**286**(1–3):97-109. DOI: 10.1016/S0048-9697(01)00967-6

[52] Capuano E, Rademaker J, van den Bijgaart H, van Ruth SM. Verification of fresh grass feeding, pasture grazing and organic farming by FTIR spectroscopy analysis of bovine milk. Food Research International. 2014;**60**:234-241. DOI: 10.1016/j.foodres.2013.12.024

[53] Bell MJ, Mereu L, Davis J. The use of mobile near-infrared spectroscopy for real-time pasture management. Frontiers in Sustainable Food Systems. 2018;**2**:76. DOI: 10.3389/fsufs.2018.00076

[54] Restaino EA, Fernández EG, La Manna A, Cozzolino D. Prediction of the nutritive value of pasture silage by near infrared spectroscopy (NIRS). Chilean Journal of Agricultural Research. 2009; **69**(4):560-566

[55] Danelli PP, Carlini P, Bernabucci U, Ronchi B. Quality evaluation of regional forage resources by means of near infrared reflectance spectroscopy. Italian Journal of Animal Science. 2016;**3**(4): 363-376. DOI: 10.4081/ijas.2004.363

[56] Catunda KL, Churchill AC, Power SA, Moore BD. Near infrared spectroscopy calibration strategies to predict multiple nutritional parameters of pasture species from different

functional groups. Journal of Near Infrared Spectroscopy. 2022;**30**(5): 254-263. DOI: 10.1101/2021.07.31.454175

[57] Parrini S, Acciaioli A, Crovetti A, Bozzi R. Use of FT-NIRS for determination of chemical components and nutritional value of natural pasture. Italian Journal of Animal Science. 2019; **17**(1):87-91. DOI: 10.1080/1828051X.2017.1345659

[58] Ribeiro MCA, Guerra GL, Cano C, Nóbrega de Carvalho L, Galbeiro S, Siqueira PR, et al. Prediction models of the nutritional quality of fresh and dry *Brachiaria brizantha* cv. Piatã grass by near-infrared spectroscopy. Journal of Applied Animal Research. 2023;**51**(1): 193-203. DOI: 10.1080/09712119.2023.2172022

[59] Smith C, Cogan N, Badenhorsh P, Spangenberg G, Smith K. Field spectroscopy to determine nutritive value parameters of individual ryegrass plants. Agronomy. 2019;**9**(6):1-14. DOI: 10.3390/agronomy9060293

[60] Wijesingha J, Astor T, Schulze-Brüninghoff D, Wengert M, Wachendorf M. Predicting forage quality of grasslands using UAV-borne imaging spectroscopy. Remote Sensing. 2020;**12**(1):126. DOI: 10.3390/rs12010126

[61] Kartal S. Comparison of semantic segmentation algorithms for the estimation of botanical composition of clover-grass pastures from RGB images. Ecological Informatics. 2021;**66**:101467. DOI: 10.1016/j.ecoinf.2021.101467

[62] Foley WJ, McElwee A, Lawler I, Aragonés L, Woolnough AP, Berding N. Ecological applications of near-infrared reflectance spectroscopy a tool for rapid, cost-effective prediction of the composition of plant and animal tissues and aspects of animal performance. Oecologia. 1998;**116**(3):293-305. DOI: 10.1007/s004420050591

[63] Verrelst J, Camps-Valls G, Muñoz J, Rivera JP, Veroustraete F, Clevers JG, et al. Optical remote sensing and the retrieval of terrestrial vegetation bio-geophysical properties- a review. Journal of Photogrammetry and Remote Sensing. 2015;**108**:273-290. DOI: 10.1016/j.isprsjprs.2015.05.005

[64] Calle A, Salvador P. Revisando el concepto de resolución en teledetección. Revista de Teledetección. 2012;**37**:74-79

[65] Geladi P. Chemometrics in spectroscopy. Part 1, classical chemometrics. Spectrochimica Acta Part B: Atomic Spectroscopy. 2003;**58**(5): 767-782. DOI: 10.1016/S0584-8547(03)00037-5

[66] Peng X, Shi T, Song A, Chen Y, Gao W. Estimating soil organic carbon using VIS/NIR spectroscopy with SVMR and SPA methods. Remote Sensing. 2014;**6**(4):2699-2717. DOI: 10.3390/rs6042699

[67] Minasny B, McBratney AB, Stockmann U, Hong SY. Cubist a regression rule approach for use in calibration of NIR spectra. In: 16th International Conference on Near Infrared Spectroscopy 2–7 June 2013; France.

[68] Morellos A, Pantazi XE, Moshou D, Alexandridis T, Whetton R, Tziotzios G, et al. Machine learning based prediction of soil total nitrogen, organic carbon and moisture content by using VIS-NIR spectroscopy. Biosystems Engineering. 2016;**152**:104-116. DOI: 10.1016/j.biosystemseng.2016.04.018

[69] Zhang D, Zhang H, Zhao Y, Chen Y, Ke C, Xu T, et al. A brief review of new data analysis methods of laser-induced

breakdown spectroscopy: Machine learning. Applied Spectroscopy Reviews. 2020;**57**(2):1-23. DOI: 10.1080/05704928.2020.1843175

[70] Qi Y, Hu D, Jiang Y, Wu Z, Zheng M, Chen EX, et al. Recent progresses in machine learning assisted Raman spectroscopy. Advanced Optical Materials. 2023;**11**(14):1-22. DOI: 10.1002/adom.202203104

[71] Shahin MA, Symons SJ. Detection of Fusarium damaged kernels in Canada Western Red Spring wheat using visible/near-infrared hyperspectral imaging and principal component analysis. Computers and Electronics in Agriculture. 2011;**75**(1):107-112. DOI: 10.1016/j.compag.2010.10.004

[72] Su WH, Bakalis S, Sun DW. Potato hierarchical clustering and doneness degree determination by near-infrared (NIR) and attenuated total reflectance mid-infrared (ATR-MIR) spectroscopy. Journal of Food Measurement and Characterization. 2019;**13**:1218-1231. DOI: 10.1007/s11694-019-00037-3

[73] Ahmadi A, Emami M, Daccache A, He L. Soil properties prediction for precision agriculture using visible and near-infrared spectroscopy: A systematic review and meta-analysis. Agronomy. 2021;**11**(433):1-14. DOI: 10.3390/agronomy11030433

[74] Chang CW, Laird DA, Mausbach MJ, Hurburgh CR. Near-infrared reflectance spectroscopy–principal components regression analyses of soil properties. Soil Science Society of America Journal. 2001;**65**(2):480. DOI: 10.2136/sssaj2001.652480x

[75] Massignani C, Vandresen BB, Marques JV, Kazama R, Osmari MP, Silva-Kazama DC. A single calibration of near-infrared spectroscopy to determine the quality of forage for multiple species. Research, Society and Development. 2021;**10**(10):1-10. DOI: 0.33448/rsd-v10i10.18990

[76] Geipel J, Bakken AK, Jørgensen M, Korsaeth A. Forage yield and quality estimation by means of UAV and hyperspectral imaging. Precision Agriculture. 2021;**22**:1437-1463. DOI: 10.1007/s11119-021-09790-2

[77] Thomson AL, Karunaratne SB, Copland A, Stayches D, Morse E, Jacobs J. Use of traditional, modern, and hybrid modeling approaches for in situ prediction of dry matter yield and nutritive characteristics of pasture using hyperspectral datasets. Animal Feed Science and Technology. 2020;**269**: 114670. DOI: 10.1016/j.anifeedsci.2020.114670

[78] Duranovich FN, Yule IJ, Lopez-Villalobos N, Shadbolt NM, Draganova I, Morris S. Using proximal hyperspectral sensing to predict herbage nutritive value for dairy farming. Agronomy. 2020;**10**(11):18-26. DOI: 10.3390/agronomy10111826

[79] Safari H, Fricke T, Wachendorf M. Determination of fiber and protein content in heterogeneous pastures using field spectroscopy and ultrasonic sward height measurements. Computers and Electronics in Agriculture. 2016;**123**: 256-263. DOI: 10.1016/j.compag.2016.03.002

[80] Berauer BJ, Wilfahrt PA, Reu B, Schuchardt MA, Garcia-Franco N, Zistl-Schlingmann M, et al. Predicting forage quality of species-rich pasture grasslands using vis-NIRS to reveal effects of management intensity and climate change. Agriculture, Ecosystems and Environment. 2020;**296**:106929. DOI: 10.1016/j.agee.2020.106929

[81] Murphy DJ, O'Brien B, O'Donovan M, Condon T, Murphy MD. A near-infrared spectroscopy calibration for the prediction of fresh grass quality on Irish pastures. Information Processing in Agriculture. 2022;**9**(2):243-253. DOI: 10.1016/j.ecoinf.2021.101467

[82] Parrini S, Staglianò N, Bozzi R, Argenti G. Can grassland chemical quality be quantified using transform near-infrared spectroscopy? Animals. 2021;**12**(1):86. DOI: 10.3390/ani12010086

Chapter 7

Cation Exchange Capacity in Grazing Systems and a Case Study for Quantification by Hyperspectral Imaging

Tatiana Rodríguez Monroy, Manuela Ortega Monsalve, Luis Fernando Galeano-Vasco, Marisol Medina-Sierra and Mario Fernando Ceron-Munoz

Abstract

This chapter provides an overview of cation exchange capacity (CEC) and its importance as an indicator of soil fertility, particularly in the assessment of grassland quality. The limitations of traditional methods are highlighted, and the need to explore more agile approaches to grassland quality assessment is emphasized. The increasing use of hyperspectral information (HSI) as an accurate tool for measuring soil properties, which promotes more effective and sustainable rangeland management, is further explored. This provides data on soil fertility and forage quality, enabling more accurate decisions. The benefits and challenges of using HSI data to estimate CEC and its potential to improve pasture and forage production will also be examined. HSI technology allows information to be collected and analyzed from reflected light at different wavelengths, providing a clear understanding of soil physical and chemical properties. In addition, a case study illustrating the estimation of CIC using hyperspectral cameras in the department of Antioquia, Colombia, is presented. The chapter emphasizes the relevance of this topic in the rangeland context and concludes with a future outlook that anticipates a change in the management and understanding of grazing systems.

Keywords: agriculture, fertility, soil, spectroscopy, supervised algorithms, remote sensing

1. Introduction

The soil is a nonrenewable natural resource that provides essential ecosystem and environmental services. It acts as a support system and nutrient provider for plant growth, supporting food production, and supplying raw materials for a wide range of human activities [1]. To assess the condition and sustainability of the soil, it is necessary to consider its quality, taking into account the integration of chemical, physical, and biological factors [2].

The physical properties of soil are fundamental to understanding its ability to store and release nutrients and to facilitate plant root development. These properties include texture (sand, silt, and clay content), structure (aggregates), porosity (spaces between particles), bulk density, and water holding capacity [3]. Soil respiration, microbial biomass, nitrogen mineralization, and earthworm density are also considered to assess the biological aspects of the soil [4]. On the other hand, soil chemical properties also influence nutrient availability and are essential for plant growth. Among the most important are soil pH, which controls nutrient availability, CEC, which indicates the soil's ability to retain nutrients, and the concentration of primary, secondary, and micronutrients [5].

In this context, soil quality serves as a tool to examine changes caused by soil management practices, such as excessive fertilization, inadequate irrigation, uncontrolled grazing, nutrient depletion through erosion, physical decomposition of aggregates due to over-tillage, loss of organic matter, salinization, and alkalinization [6]. Soil fertility management is one of the most critical decisions as nutrients often limit plant growth and animal performance [5]. Nutrient balances in the grass are crucial in defining fertilization needs when the nutrient balance is negative and in reducing the purchase of off-farm inputs [7]. Therefore, the objective of this chapter is to analyze CEC in grass production and to present a case study for its determination using HSI technology.

2. CEC in grass production and the use of spectra in its estimation

2.1 CEC as an indicator of soil fertility

CEC is a soil property that describes its ability to provide nutrients in the soil solution, making them available for plant uptake. It is a key indicator for assessing soil fertility [8] as it determines the soil's capacity to retain and supply cations essential for healthy plant growth and development. CEC is defined as the sum of exchangeable cations in the soil, including calcium (Ca^{2+}), magnesium (Mg^{2+}), potassium (K^{+}), sodium (Na^{+}), hydrogen (H^{+}), aluminum (Al^{3+}), iron (Fe^{2+}), manganese (Mn^{2+}), zinc (Zn^{2+}), and copper (Cu^{2+}), expressed in $cmol^{+}kg^{-1}$ or meq/100 g (milliequivalents per 100 grams of soil) [9].

The term "cation exchange" refers to the process by which cations are exchanged in the soil solution, and subsequently absorbed by plant roots. The presence of clay minerals and organic matter (OM) in the soil enhances the amount of cations in the soil solution due to their negative charge, which attracts positively charged ions (cations) to their surfaces through electrostatic forces. As a result, the cations remain within the root zone of the soil (**Figure 1**).

In soil, OM can exhibit CEC by weight that is 4–50 times higher than that of clay. Unlike clay minerals, OM has a distinct negative charge source. This negative charge results from the dissociation of organic acids within OM, resulting in a net negative charge that is balanced by the presence of cations in the soil. This negative charge on OM is referred to as pH-dependent CEC as the dissociation of organic acids is affected by soil pH. Consequently, the actual soil CEC varies as a function of soil pH. For example, a neutral soil with pH 7 containing the same amount and type of OM will have a higher CEC than a soil with a lower pH, such as pH 5 [9]. On the other hand, clay has a high capacity to attract and retain cations due to its chemical structure. The

Figure 1.
Cation exchange in the root zone of a grass. Source: Own elaboration.

ratio of CEC to clay content in weight percent can vary significantly because clay minerals exhibit different CECs due to variations in their structure and chemical composition. The proportion of exchangeable cations also varies among different clay minerals. For instance, montmorillonite clay has a higher CEC, heavily weathered kaolinite clay has a lower CEC, and slightly higher CEC is found in the less weathered illite clay [10]. In this context, the estimation of CEC in livestock production systems becomes crucial as it identifies the soil's ability to retain nutrients and provide a favorable environment for pasture growth. By utilizing soil testing techniques, producers can determine the CEC and take appropriate measures to enhance it. This involves the proper application of amendments and fertilizers, adjusting soil pH, and adopting grazing management practices that prevent soil compaction and promote MO incorporation [11]. Improving the soil's CEC ensures the adequate availability of essential nutrients for pasture growth, leading to higher livestock productivity. Moreover, enhancing soil conditions contributes to the conservation of ecosystem structure and health, promoting long-term sustainable management.

2.2 Methods used for the determination of CEC

To quantify CEC and evaluate soil fertility, various direct or conventional methods, as well as indirect or addition methods, have been established. Direct methods involve determining CEC as a single measure by saturating the soil exchange sites with a solution of specific cations, allowing for unique and quantitative measurements. These methods enable CEC determination at different soil pH levels using unbuffered reagents such as KCl, NH_4Cl, and organometallic cations, as well as in a buffered medium to eliminate pH variation in measurements and express all results on the same basis [12]. On the other hand, indirect methods use related parameters, such as exchangeable bases and exchangeable acidity and employ equations to determine CEC. This involves summing up cations and displacing exchangeable cations in the soil using a saturating salt solution, such as ammonium acetate [13].

Currently, the standard reference method most commonly used by laboratories to determine CIC is to saturate the exchange complex with ammonium acetate (1 N at pH 7) for both acidic and alkaline soils. This method uses an ammonium acetate solution to displace the exchangeable cations present in the soil and measures the concentration in the solution to determine the CEC. Another method is the calculation with the sum of exchange bases (Ca^{+2}, Mg^{+2}, K^{+1}, and Na^{+1}). For this process, extraction of 30 to 60 ml of ammonium acetate and 50 ml of 96% ethyl alcohol is required, and for quantification, 50 ml of 10% sodium chloride, 20 ml of formaldehyde, phenolphthalein, and sodium hydroxide are used [14].

The determination of CEC in the laboratory, through direct and indirect methods, presents certain limitations. The preparation of samples, chemical analysis, and interpretation of results requires considerable time, qualified personnel, and specialized equipment. However, a promising alternative for CEC determination is HSI reflectance spectroscopy. This nondestructive technique enables a rapid and efficient analysis of soil, providing detailed information about its properties. By utilizing HSI images, soil reflectance is captured across multiple spectral bands, allowing for the detection of interactions among soil chemical elements. This novel approach has the potential to overcome some of the limitations associated with traditional laboratory methods, facilitating a more precise and detailed assessment of CEC in different soil areas [15].

2.3 Spectroscopy, an alternative to assess soil fertility

The need for faster and more cost-effective analysis has led to the widespread use of infrared spectroscopy. Spectroscopy uses the interaction of light with soil components to provide valuable information about soil properties. The technique is based on measuring the reflection, absorption, and emission of light at different wavelengths to characterize the chemical composition and physical structure of the soil [16]. By analyzing the amount of reflected, absorbed, and transmitted light in each spectral band, detailed data on the different elements and compounds present in the soil can be obtained. This allows a rapid and nondestructive assessment of the chemical and mineralogical composition of the soil, including an estimate of the CEC. This approach reduces the need for traditional laboratory analysis and provides faster results, which can be particularly useful in situations where frequent monitoring of CEC is required over large areas or in resource-limited conditions [17]. It is an emerging technology that successfully predicts the physicochemical properties of soils [15], including CEC, pH, clay content, carbon content, total nitrogen content, and other elements, by correlating spectral data extracted from images with their chemical and physical concentrations [18–20].

This technique is based on the fact that materials reflect electromagnetic energy in the form of different patterns and wavelengths due to their chemical composition, physical structure, and surface properties. HSI captures the radiation reflected from objects in many very narrow spectral bands, creating large data cubes per pixel [21]. The data cubes contain the radiance received by the sensor in a particular band of the spectrum, corresponding to the size of the pixel, which is the smallest visual unit that appears in the image. Each pixel is defined by an integer number known as the digital level (ND). In this sense, the information of an image can be represented as a three-dimensional numerical matrix since it has spatial information on its X and Y axes, corresponding to the geographical coordinates of the image, and spectral information on the Z axis. Considering that HSI images contain a large amount of data, handling, storage, and processing is challenging due to the high spectral variability and correlation in the data. The analysis of large amounts of data involves the following phases:

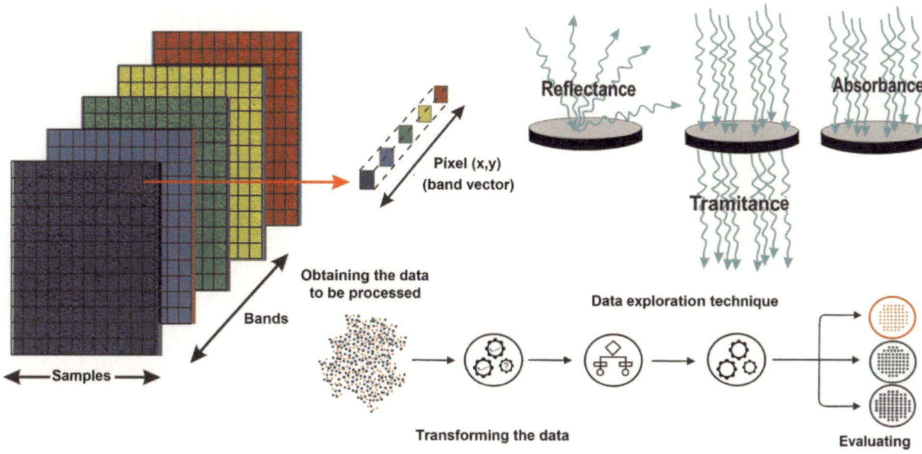

Figure 2.
Hyperspectral information from a hyperspectral image. Adapted from [23].

obtaining the data to be processed, transforming the data so that it can be used, applying the data exploration technique, and evaluating the results obtained [22], as shown in **Figure 2**.

Data transformations are applied to HSI data to reduce noise, improve quality, and enable modeling. Some transformations used for noise removal are the Savitzky-Golay (SG) filter, also known as the digital smoothing polynomial, which reduces the effect of noise without causing much distortion in the spectrum, especially in the width and height of the bands. It is also simple and efficient and involves only a linear convolution with a set of filter coefficients [24]. Another technique used is standard normal variation (SNV) [25], which highlights important patterns and relationships between different bands. On the other hand, the detrending technique (DT) corrects the trend of the data [26].

Similarly, several data exploration techniques have been used to relate the spectra to soil properties and have been used to relate the spectra to soil properties. Among the most widely used are partial least squares regression (PLSR), which can handle large and noisy data sets, and the support vector machine (SVM) method, which is characterized by its ability to generate robust models with few training samples. Other methods, such as artificial neural networks (ANN), are also used in HSI analysis. ANNs consist of an input layer, one or more hidden neuron layers, and an output layer. Decision trees and random forests (RF), which are built from rules, have bifurcations or branches that depend on a condition based on linear regression. These algorithms have been widely used in the scientific literature for the prediction of various soil characteristics using spectral or multidimensional data [27]. Therefore, the aim of this paper is to evaluate supervised learning algorithms in soil CIC estimation from HSI images.

3. Case study

3.1 Generalities

Livestock production is a rapidly growing sector in Colombia. At present, the country's cattle population is spread over 620.807 farms, with a total of 29.642.539

animals, an increase of 1.2% compared to 2022 [28]. This economic activity covers 34 million hectares of land [29], divided into three production systems: dairy, beef, and dual purpose, of which 70% is managed under extensive production systems, with an average stocking density of 0.9 animals per hectare [30].

Livestock diets are based on pasture and forage, with occasional use of concentrates [29]. Although precise data on the range of forage species used in livestock production are not available, it is widely recognized that kikuyu grass (*Cenchrus clandestinus* (Hochst. ex Chiov.) Morrone) plays a crucial role as a forage base in specialized dairy production in high tropical climates [31, 32]. Kikuyu grass is particularly dominant in these areas and provides a constant and abundant source of forage for cattle [33]. In the case of cattle in the mid and low tropics, it has been observed that the most commonly used forage species are star grass (*Cynodon nlemfuensis*), various species of *Brachiaria* (such as *B. decumbens*, *B. humidicola* and *B. brizantha*) and some native species [34].

Livestock productivity is closely linked to the ability of producers to manage their pastures effectively. Proper management involves the accurate and timely application of nutrients necessary for plant growth, which promotes early pasture recovery and maintains sustainable forage production throughout the year [35]. In this sense, a fundamental strategy is to ensure that the soil has the optimum physical and chemical characteristics for adequate plant development. This ensures an environment conducive to the growth of healthy and high-quality grasses.

In the search for faster and more cost-effective alternatives for estimating CEC and assessing soil quality, infrared spectroscopy is emerging as a viable option. This technique provides rapid and nondestructive measurements of soil chemical and mineralogical composition, including CEC estimation, by analyzing the reflection, absorption, and transmission of light at different wavelengths. By using spectroscopy, producers can obtain detailed information on nutrient availability and soil characteristics, enabling them to make informed decisions on nutrient application and pasture management to maximize livestock productivity.

The study area for the estimation of CEC by infrared spectroscopy was located in the department of Antioquia, Colombia, covering the nine subregions and 96 municipalities of the department, as shown in **Figure 3**. The information was obtained from 1997 soil samples collected between November 2020 and November 2022 in pastoral and cocoa production systems. The altitudes of the sampled areas ranged from 0 to 2900 masl and were characterized by a high spatial and temporal variability.

3.2 Methodology

To determine CEC in soil, 1997 samples were analyzed according to the Colombian technical standards NTC 5667:2017 for soil sampling in the field [36] and NTC 5805:2003 for sample preparation for chemical analysis [37]. Laboratory analysis of CEC was carried out at the Colombian Agricultural Research Corporation AGROSAVIA using the indirect method of sum of bases [38].

To record and extract HSI information, HSI images were taken by drying the soil samples at 40°C for 48 hours in a forced-air oven and sieving them at 2 mm. The images were taken using two cameras, the Baldur S-384 N or SWIR (short wave infrared) and the Baldur V-1024 N or VNIR (visible and near infrared). The SWIR camera had a spectral range of 951.61–2517.86 nm, 288 spectral bands, 384 spatial pixels, and a 30 cm lens, while the VNIR camera had a spectral range of 485.14–955.65 nm, 88 spectral bands, 1024 spatial pixels, and a 30 cm lens. Reflectance (R)

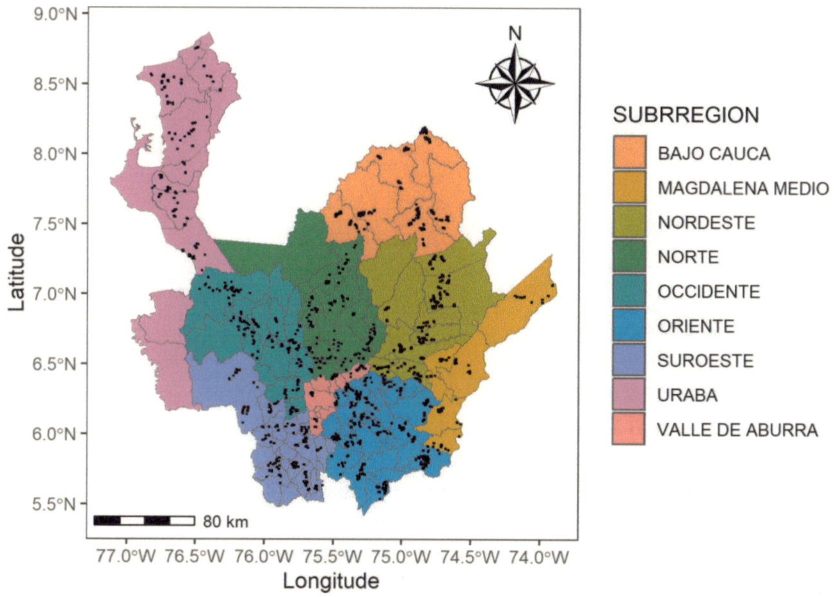

Figure 3.
Location of soil samples collected in the 96 municipalities of the nine subregions of the department of Antioquia.

values were then extracted from the SWIR and VNIR cameras in a region of interest located in the center of the image to avoid pixel errors. Digital masking was applied to pixels with R values greater than 0.2 and less than 0.9 to remove saturated and shaded pixels. This process was performed iteratively using the SpectralPy [39], Spectral [40], and NumPy [41] libraries of the Python 3.8.2 programming language [42].

The overlapping bands detected in the transition region between 950 and 960 nm were removed using the Spectrolab library created by [43] in the R-Project programming language [44]. Band 955 was removed so that the spectra recorded with the VNIR camera ended at band 951, while the SWIR camera started at band 957. Finally, the hyperspectral band database was obtained.

To build the models, the CEC variable was used in its original form and transformed using the Box-Cox method with a value of $\lambda = -0.1547$ and the square root transformation. In addition, the reflectance values were converted to absorbance (ABS) using (Eq. (1)).

$$ABS = \log_{10} \frac{1}{R} \quad (1)$$

In addition, SG, SNV, and DT transformations were applied using the prospectr library [45].

PLSR algorithms were applied to these combinations using 5, 10, 15, 20, 25, and 28 components. Each component was constructed by linearly combining the original variables with the intention of maximizing the covariance between the predictor variables and the response variable. SVM models with linear and polynomial kernels were used, with model building costs of 500 and 100, respectively. The data set was divided into training data (train) with 75% of the data and test data (test) with 25% of the data.

To assess the performance and fit of the models, the coefficient of determination (R^2) was used for both the training and test data. In addition, the residual prediction variance (RPD) was calculated. R^2 is obtained by squaring the correlation between the predicted and actual values (Eq. (2)) and provides a measure of how well the model fits the data.

$$R^2 = 1 - \frac{\sum_{i=1}^{n} (y_i - \hat{y}_i)^2}{\sum_{i=1}^{n} (y_i - \bar{y})^2} \tag{2}$$

where:
n : number of observations
y_i : real values
$\hat{y}_i$: values predicted by the model y
$\hat{y}$: mean of the real values

Eq. (3) defines the ratio of performance to deviation (RPD), which is the ratio of the standard deviation of the reference values (observations) to the standard deviation of the differences between the observations and the model predictions. The RPD is a measure used to assess the performance of a predictive model by comparing the accuracy of the reference values with the variability of the model's predictions [46].

$$RPD = \frac{\sqrt{\frac{1}{n-1} \sum_{i=1}^{n} (y_i - \bar{y})^2}}{\sqrt{\frac{1}{n} \sum_{i=1}^{n} (y_i - \hat{y}_i)^2}} \tag{3}$$

3.3 Results and discussion

The CEC values the soils in the department of Antioquia show a wide variability, ranging from a minimum of 0.21 to a maximum of 61.8 cmol (+)/kg. The mean CEC obtained was 8.62 ± 3.33 cmol (+)/kg, with a coefficient of variation of 123.86 cmol (+)/kg. The mode was 0.62 and the median was 33.3 cmol (+)/kg. These results indicate significant variation in CEC across the department, with most samples having low levels of CEC. These results are consistent with previous studies carried out in the northern, northeastern, and Urabá regions of Antioquia [47, 48].

The CEC variable was transformed using the Box-Cox method, resulting in transformed data with a minimum dimension of -1.76 and a maximum dimension of 3.04. The mean obtained after transformation was 1.09 ± 0.99 cmol (+)/kg, with a mode of -0.49 and a median of 1.09 cmol (+)/kg. We also performed the transformation using the square root of CEC, which gave values ranging from a minimum of 0.45 to a maximum of 7.86. The resulting mean was 1.82 ± 1.41 cmol (+)/kg, with a mode of 0.78 and a median of 1.09 cmol (+)/kg. The transformation successfully reduced the scale of the data, as shown in **Figure 4**.

The spectral reflectance of the soil increased as the wavelength increased and remained consistent across different soil samples. Figure illustrates the pattern of the average curves from various samples, displaying several absorption valleys. A reflection peak was observed between 950 and 957 nm, caused by the overlapping noise from VNIR and SWIR sensors, which aligns with the findings described by [49].

The spectral reflectance of the soil increased with wavelength in the visible, near-infrared (NIR) and mid-infrared (MIR) regions, and this behavior was consistent across the different soil samples. A prominent reflection peak was observed between 950 and 957 nm, attributed to the overlapping noise of the VNIR and SWIR sensors,

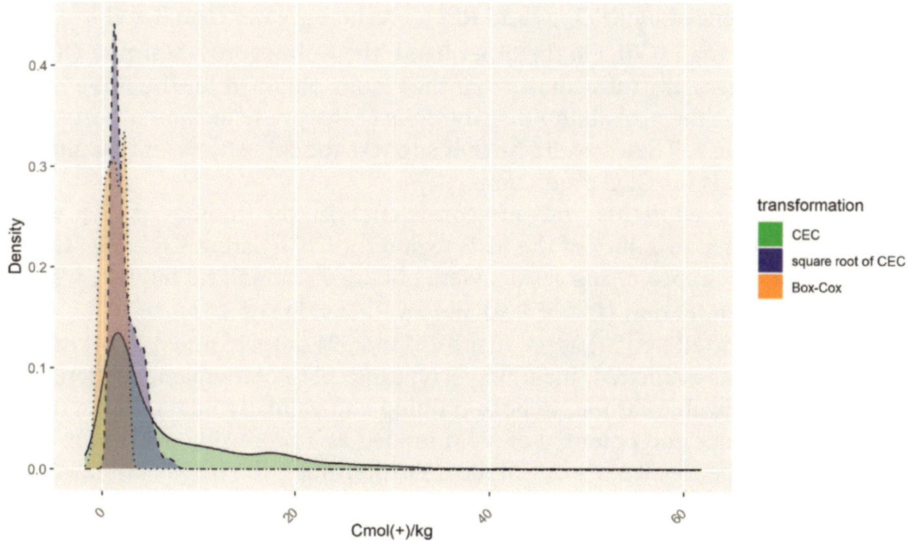

Figure 4.
Diagram of soil CEC distribution, square root, and Box-Cox.

which is consistent with the results described by [49]. In addition, the **Figure 5** illustrates the average curve patterns of the analyzed samples, showing three absorption valleys in the NIR and MIR regions. These observations are consistent with those reported by [50], suggesting that these valleys correspond to the hydroxyl and clay absorption bands of the soils.

A total of 144 PLSR models and 97 SVM models were evaluated for CEC prediction. The PLSR models showed promising results, achieving a RPD_{train} greater than 2, indicating their ability to explain data variability. However, the SVM models outperformed the PLSR models, showing even better performance in explaining data

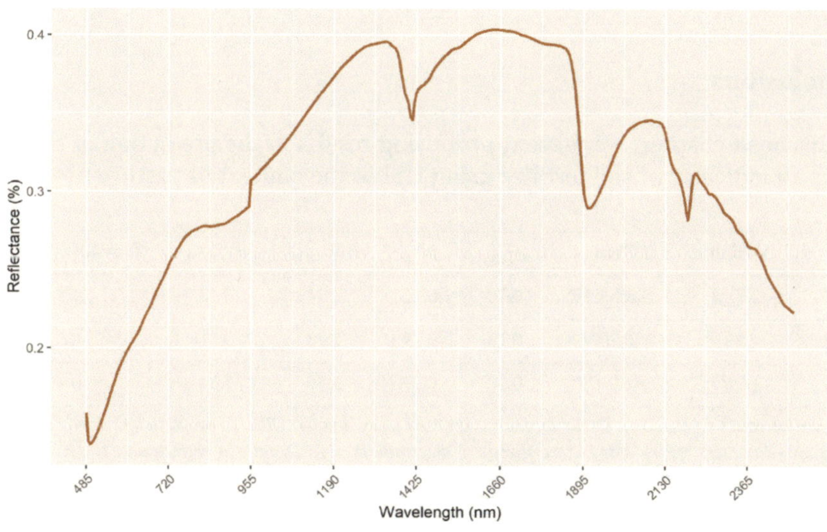

Figure 5.
The spectral signature of the soil.

variability, as evidenced by RPD_{train} and RPD_{test} values greater than 3 and R^2_{train} and R^2_{test} values greater than 0.70. On the other hand, the Kolmogorov-Smirnov (KS) test to assess the goodness of fit test shows that the results obtained for the three best models indicate that the probability distributions of the predicted and observed data adequately fit **Table 1**. These results further support the robustness and quality of the selected SVM models for CEC prediction.

The results obtained in this study are consistent with the findings of [51], who evaluated the predictive ability of the MIR region for CIC (Cation Exchange Capacity). In their research, promising results were obtained using PLSR models, with a coefficient of determination (R^2) of 0.92 and an RPD value of 3.5.

The results reported by [52] agree that SVM models outperformed PLSR models for all soil properties evaluated, including clay, sand, pH, total organic carbon, and permanganate oxidizable carbon, in both training and validation data. These results confirm the relevance and potential of SVM models as a promising option for predicting soil properties from spectral data. Furthermore, the results obtained are in line with the previous study by [53], where they determined the CEC of 142 samples using a spectral range from 350 to 2500 nm. They found a PLSR model with $R^2 = 0.76$ in the training data and $R^2 = 0.72$ in the test data.

On the other hand, it is observed that the models perform better when working with the CEC variable transformed with *sqrtCEC*. As for the HSI data, high performance is obtained with the SNV and DT transformations applied. These results are in line with the results of a study carried out by [41], which showed that preprocessing methods for hyperspectral data improve the accuracy of the evaluated models.

Figure 6 shows a scatter plot comparing the predicted and observed values of the SVM model. The blue dots represent the observed values, while the red dots represent the predicted values. It can be seen that the red dots follow a trend close to the diagonal line, suggesting that the SVM model has effectively captured the variability in the data. It is also important to note that applying the KS test (p-value) to the data yielded values greater than 0.05, indicating that there is no significant difference between the observed and predicted values.

4. Conclusions

In this book chapter, we present promising results in the prediction of the CEC index as an indicator of soil fertility using HSI in the context of pasture production.

Alg.	Variable	Trans	R^2_{train}	R^2_{test}	RPDtrain	RPDtest	Test KS (p-value)
SVM	$\sqrt{CEC}$	ABS-SNV	0.78	0.79	3.44	3.52	0.44
SVM	$\sqrt{CEC}$	R-SNV	0.78	0.79	3.43	3.51	0.71
SVM	$\sqrt{CEC}$	ABS-DT	0.78	0.77	3.40	3.44	0.46

Alg., Algorithm used for modeling; Variable, conversion method used in the CIC; Trans, data transformation method; train, training data; test, testing data; R^2, coefficient of determination; RPD, ratio of performance to deviation; TestKS($p - value$), p-value of the Kolmogorov–Smirnov test; SVM, support vector machines; CEC, cationic interaction capacity; ABS, absorbance; R, reflectance; SNV, standard normal variation; DT, detrending technique.

Table 1.
Results of the five best performing models for predicting cation exchange capacity in the soil.

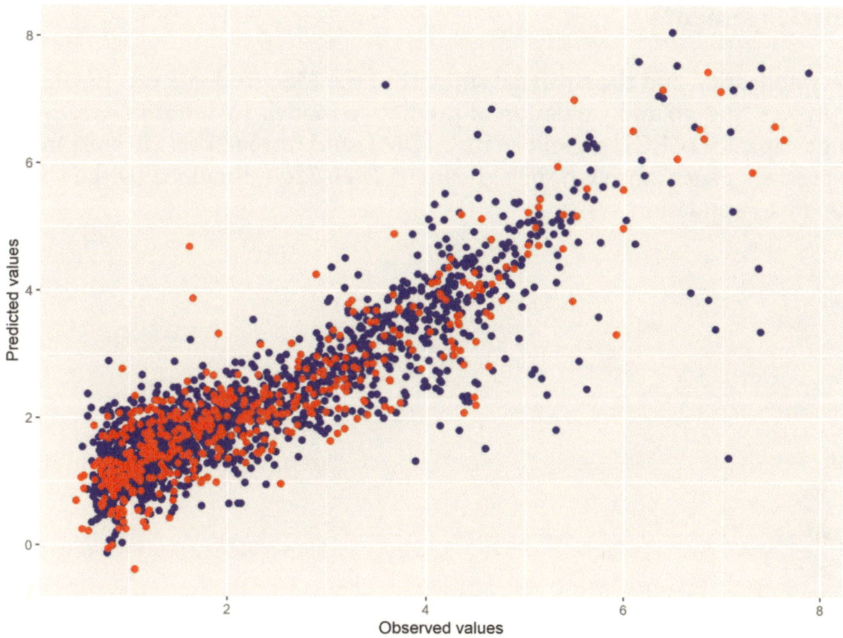

Figure 6.
Observed and predicted values from the SVM model.

The application of HSI technology has shown significant potential for assessing optimal soil conditions and improving management strategies in pasture and forage production. However, it is essential to recognize and address the challenges associated with the use of HSI data in the analysis of soil physicochemical parameters. HSI techniques represent an interdisciplinary field that incorporates and adapts different concepts, approaches, and algorithms. Several signals underline the growing importance of HSI technology in remote sensing. One such indicator is the continuous increase in the number of HSI sensors, supported by the many applications that rely on this technology. In addition, the increase in the number of scientific publications also supports this growth. It is important to note that the methods used to analyze remotely sensed HSI data are not always straightforward. Key challenges include the processing of large amounts of information, high dimensionality, and the need for specialized techniques to extract meaningful patterns and insights. In addition, rigorous data pre-processing and calibration processes are essential to ensure the accuracy and reliability of predictions. One of the most obvious barriers is the management of large data archives, which is hampered by the lack of specialized hardware [27]. As the field of HSI analysis continues to develop, it is imperative that researchers and practitioners develop creative strategies to overcome these barriers and maximize the benefits of HSI technology in soil fertility improvement and pasture production management. Overcoming these challenges will unlock the true potential of HSI technology and usher in a transformative era in soil-related agricultural research and applications. By fully embracing HSI technology and actively addressing its challenges, there is an opportunity to unlock the full potential of hyperspectral data and stimulate positive changes in soil fertility, rangeland productivity, and sustainable agricultural practices.

Acknowledgements

This publication and the financial support of the MSc students was possible thanks to the project "design and validation of predictive models to determine cation exchange capacity (CEC), organic matter (OM), and nitrogen (N) in soils from hyperspectral images" through the agreement 2022-7204, financed by the University of Antioquia Foundation.

Author details

Tatiana Rodríguez Monroy[1*], Manuela Ortega Monsalve[1], Luis Fernando Galeano-Vasco[2], Marisol Medina-Sierra[2] and Mario Fernando Ceron-Munoz[2]

1 Animal Sciences, GAMMA Research Group, Faculty of Agricultural Sciences, Antioquia University, Medellin, Antioquia, Colombia

2 GAMMA Research Group, Faculty of Agricultural Sciences, Antioquia University, Medellin, Antioquia, Colombia

*Address all correspondence to: tatiana.rodriguezm@udea.edu.co

References

[1] El Burbano-Orjuela H. suelo y su relación con los servicios ecosistémicos y la seguridad alimentaria. Revista de Ciencias Agrícolas. 2015;**33**(2):117-124. DOI: 10.22267/rcia.163302.58

[2] de la Rosa D, Sobral R. Land Use and Soil Resources Chapter Soil Quality and Methods for its Assessment. Dordrecht: Springer; 2008. pp. 167-200. DOI: 10.1007/978-1-4020-6778-59

[3] Cruz A, Barra J, del Castillo RF, Gutiérrez C. La calidad del suelo y sus indicadores. Revista Científica y Técnica de Ecología y Medio Ambiente Ecosistemas. 2004;**13**(2):90-97

[4] Bünemann EK, Bongiorno G, Bai Z, Creamer RE, De Deyn G, de Goede R, et al. Soil quality – A critical review. Soil Biology and Biochemistry. 2018;**120**: 105-125. DOI: 10.1016/j.soilbio.2018.01.030

[5] Silveira ML, Kohmann MM. Maintaining soil fertility and health for sustainable pastures. In: Chapter 3: Management Strategies for Sustainable Cattle Production in Southern Pastures. Florida, United States: Academic Press; 2020. pp. 35-58. DOI: 10.1016/b978-0-12-814474-9.00003-7

[6] Vasu D, Tiwary P, Chandran P, Singh SK. Nutrient dynamics for sustainable crop production. In: Meena R, editor. Chapter: Nutrient Dynamics for Sustainable Crop Production. Singapur: Springer; 2019. pp. 41-66. DOI: 10.1007/978-981-13-8660-22

[7] Dubeux CB, Sollenberger LE. Management strategies for sustainable pastures. In: Chapter 4 - Nutrient Cycling in Grazed Pastures. Florida, United States: Academic Press; 2020. pp. 59-75. DOI: 10.1016/B978-0-12-814474-9.00004-9

[8] García Y, Ramírez S, Sánchez S. Soil quality indicators: A new way to evaluate this resourceIndicadores de la calidad de los suelos: una nueva manera de evaluar este recurso. Pastos y Forrajes. 2012; **35**(2):125-137

[9] Martel YA, De Kipe CR, Laverdiere MR. Cation-exchange capacity of clay-rich soils in relation to organic matter, mineral composition, and surface area. Soil Science Society of America Journal. 1978;**42**(5):764-767

[10] Tan KH, Dowling PS. Effect of organic matter on CEC due to permanent and variable charges in selected temperate region soils. Geoderma. 1984; **32**(2):89-101. DOI: 10.1016/0016-7061(84)90065-X

[11] Noreña GJ. Indicadores de calidad del suelo relacionados con la degradación de la pastura y el rendimiento forrajero del pasto Kikuyo *Cenchrus clandestinus* (Hochst. ex Chiov.) Morrone. Medellín, Colombia: Universidad Nacional de Colombia; 2022

[12] Castillo IA. Comparación de métodos para determinar la capacidad de intercambio catiónico en diez suelos venezolanos. Maracay, Venezuela: Universidad Central de Venezuela; 2019. pp. 27-29

[13] Norman AG, Chapman HD. Methods of Soil Analysis: Part 2 Chemical and Microbiological Properties Chapter 58 Total Exchangeable Bases, Contribution from the Department of Soils and Plant Nutrition. Riverside, CA: University of California; 1995. pp. 902-904. DOI: 10.2134/agromonogr9.2.c7

[14] Instituto Colombiano de Normas Técnicas y Certificación ICONTEC. Gestión ambiental. calidad del suelo.

Pretratamiento de las muestras de suelo para análisis fisicoquímicos NTC-150 11464. Bogotá, DC; 1995

[15] Vargas-Zapata M, Medina-Sierra M, Galeano L, Cerón-Muñoz M. Algoritmos de aprendizaje de máquina para la predicción de propiedades fisicoquímicas del suelo mediante información espectral: una revisión sistemática. Revista de Investigación, Desarrollo e Innovación. 2022;**12**(1):107-120. DOI: 10.19053/20278306.v12. n1.2022.1421

[16] Ferrero A, Frisvad JR, Simonot L, Santafé P, Schirmacher A, Campos J, et al. Fundamental scattering quantities for the determination of reflectance and transmittance. 2021;**29**(1):219-231. DOI: 10.1364/OE.410225

[17] Parody-Zuleta AM, Zapata-Sanjuán ED. Agricultura de precisión en Colombia utilizando teledetección de alta resolución. Suelos Ecuatoriales. 2018;**48**(1):41-49

[18] Lin L, Gao Z, Liu X. Estimation of soil total nitrogen using the synthetic color learning machine (sclm) method and hyperspectral data. Geoderma. 2020;**380**(15):1-10. DOI: 10.1016/j.geoderma. 2020.114664

[19] Guo L, Zhang H, Shi T, Chen Y, Jiang Q, Linderman M. Prediction of soil organic carbon stock by laboratory spectral data and airborne hyperspectral images. Geoderma. 2019;**337**:32-41. DOI: 10.1016/j.geoderma.2018.09.003

[20] Liu L, Ji M, Buchroithner M. Transfer learning for soil spectroscopy based on convolutional neural networks and its application in soil clay content mapping using hyperspectral imagery. Sensors. 2018;**18**(9):1-18. DOI: 10.3390/s18093169

[21] Paoletti ME, Haut JM, Plaza J, Plaza A. Estudio comparativo de técnicas de clasificación de imágenes hiperespectrales. Revista Iberoamericana de Automática e Informática Industrial. 2019;**16**(2):129-137. DOI: 10.4995/riai.2019.11078

[22] Merlino H, Britos PV, Lerache JS, Diez E, García-Martínez R. Un Método de Transformación de Datos Orientado al Uso de Explotación de Información. In: XI Congreso Argentino de Ciencias de la Computación. Buenos Aires, Argentina; 2005

[23] Chuvieco E. Fundamentos de teledetección espacial. Estudios Geográficos. 1991;**52**(203):371

[24] Beitollahi M, Hosseini SA. Using savitsky-golay smoothing filter in hyperspectral data compression by curve fitting. In: 26th Iranian Conference on Electrical Engineering. Mashhad, Irán; 2018. pp. 452-457. DOI: 10.1109/ICEE.2018.8472702

[25] Guo Q, Wu W, Massart D. The robust normal variate transform for pattern recognition with near-infrared data. Analytica Chimica Acta. 1999;**382**(1–2):87-103. DOI: 10.1016/S0003-2670(98)00737-5

[26] Nofrizal AY, Sonobe R, Yamashita H, Mihara M, Morita A, Ikka. Utilization of Hyperspectral Data and Machine Learning Algrorithms for Estimating Chlorophyll Contents in Wasabi Leaves. Vietnam: Universidad de Can Tho, ciudad de Can Tho; 2021

[27] Bioucas-Dias JM et al. Hyperspectral remote sensing data analysis and future challenges. IEEE Geoscience and Remote Sensing Magazine. 2013;**1**(2):6-36. DOI: 10.1109/MGRS.2013.2244672

[28] Instituto Colombiano Agropecuario—ICA. Censos Pecuarios

Nacionales, Censo bovino en Colombia. 2023

[29] Bravo-Parra AM. Cadenas sostenibles ante un clima cambiante. La ganadería en Colombia. Bonn (Alemania). Deutsche Gesellschaft für Internationale Zusammenarbeit (GIZ) GmbH. 2021;**2021**:16-21

[30] Mahecha L. Importancia de los sistemas silvopastoriles y principales limitantes para su implementación en la ganadería colombiana. Revista Colombiana de Ciencias Pecuarias. 2003; **16**(1):11-18

[31] Mojica JE, Castro E, León J, Cárdenas EA, Pabón ML, Carulla JE. Efecto de la oferta de pasto kikuyo *(Pennisetum clandestinum)* sobre la producción y calidad composicional de la leche bovina. Livestock Research for Rural Development. 2009;**21**(1):1-12

[32] Vargas M, Sierra A, Mancipe M, Avellaneda A. El kikuyo, una gramínea presente en los sistemas de rumiantes en trópico alto colombiano. CES Medicina Veterinaria y Zootecnia. 2018;**13**(2): 137-156. DOI: 10.21615/cesmvz.13.2.4

[33] Arango-Gaviria J, Echeverri-Zuluaga J, Lopez-Herrera AC. Diversity Kikuyu grass (*Cenchrus clandestinus*): A review. Journal of Engineering Sciences. 2019; **24**(2):81-88. DOI: 10.22463/0122820X. 1834

[34] Tapasco J, Martínez J, Calderón S, Romero G, Ordóñez DA, Sánchez-Aragón L, et al. Impactos económicos del cambio climático en Colombia: Sector Ganadero. Washington DC: Banco Interamericano de Desarrollo; 2015

[35] Motta-Delgado PA, Martínez O, Rojas-Vargas EP. Indicadores asociados a la sostenibilidad de pasturas: una revisión. Ciencia y Tecnología Agropecuaria. 2019;**20**(2):387-430. DOI: 10.21930/rcta.vol20num2art:1464

[36] Instituto Colombiano de Normas Técnicas y Certificación ICONTEC. NTC 5805:2003: Suelos. preparación de muestras para análisis químico. Norma Técnica Colombiana. 2003

[37] Instituto Colombiano de Normas Técnicas y Certificación ICONTEC. NTC 5667:2017: Suelos. toma de muestras en campo. Norma Técnica Colombiana. 2017

[38] Corporación colombiana de investigación agropecuaria (AGROSAVIA). Informe de analisis químico: Capacidad de intercambio catiónico. Technicalities 2022

[39] Piland R, Frantz D. Spectralpy: A Python Package for Hyperspectral Data Analysis. 2019. Available from: https://github.com/spectralpy/spectralpy

[40] Carriere S, Drumm I. Spectral: Python Package for Hyperspectral Data Analysis. 2021. Available from: https://github.com/spectral/spectral

[41] van der Walt S, Colbert SC, Varoquaux G. Numpy: The Fundamental Package for Scientific Computing with Python. 2011. Available from: https://numpy.org [Accessed: April 7, 2023]

[42] Van Rossum G, Drake FL. Introduction to Python 3: Python. Godalming, England: CreateSpace; 2009

[43] Maughan G, Su J, Ayers L. Spectralab: A Suite of Tools for Spectral Data Analysis in R. 2015. Available from: https://cran.r-project.org/web/packages/SpectraLab/index.html

[44] R Core Team R. A Language and Environment for Statistical Computing.

Vienna, Austria: R Foundation for Statistical Computing; 2023

[45] Malcolm EB, Peter JT, Valkenburgh, BHJ. Tools for Spectroscopy in R. 2019. Available from: https://CRAN.R-project.org/package=prospectr.rpackageversion1.1.7

[46] Wadoux AMC, Malone B, Minasny B, Fajardo M, McBratney AB. Soil Spectral Inference with r. Berlin/Heidelberg, Germany: Springer; 2021. DOI: 10.1007/978-3-030-64896-1

[47] Gordon RAG, Albanes EYA, Sánchez DAC. Evaluación de la calidad química del suelo en agroecosistemas cacaoteros de la subregión del nordeste y Urabá antioqueño. Revista Colombiana de Investigaciones Agroindustriales. 2018;**5**(1):41-52

[48] Sierra-Mediana M, Duque JA, Cerón-Muñoz M. Descripción de las relaciones entre parámetros químicos en suelos de fincas lecheras del norte de antioquia, Colombia. Investigación Ganadera para el Desarrollo Rural. 2019; **31**(7):1-12. DOI: 10.23850/24220582.730

[49] Jin X, Du J, Liu H, Wang Z, Song K. Remote estimation of soil organic matter contentin the sanjiang plain, northest China: The optimal band algorithm versus the GRA-ANN model. Agricultural and Forest Meteorology. 2016;**218–219**:250-260. DOI: 10.1016/j.agrformet.2015.12.062

[50] Streck NA, Rundquist D, Connot J. Spectral signature of selected soils. Revista Brasileira de Agrometeorologia, Santa Maria. 2003;**11**(1):181-184

[51] Lelago A, Bibiso M. Performance of mid infrared spectroscopy to predict nutrients for agricultural soils in selected areas of Ethiopia. Heliyon. 2022;**8**(3):1-6. DOI: 10.1016/j.heliyon.2022.e09050

[52] Deiss L, Margenot AJ, Culman SW, Demyan MS. Tuning support vector machines regression models improves prediction accuracy of soil properties in MIR spectroscopy. Geoderma. 2020;**365**: 114227. DOI: 10.1016/j.geoderma.2020.114227

[53] Wan M, Hu W, Qu M, Li W, Zhang C, Kang J, et al. Rapid estimation of soil cation exchange capacity through sensor data fusion of portable XRF spectrometry and Vis-NIR spectroscopy. Geoderma. 2020;**353**:1-8. DOI: 10.1016/j.geoderma.2019.114163

Chapter 8

Impact of Agro-Pastoralism on Grasslands in Serengeti and Ugalla Ecosystems, Tanzania

Pius Yoram Kavana, Bukombe John Kija, Emmanuel Pagiti Reuben, Ally Kiyenze Nkwabi, Baraka Naftal Mbwambo, Simula Peres Maijo, Selemani Rehani Moshi, Shabani Matwili, Victor Alexander Kakengi and Stephen Justice Nindi

Abstract

This chapter delves into the intricate relationship between agro-pastoralism and grassland ecosystems in Tanzania's Western Serengeti and Ugalla Ecosystems. Despite the acknowledged contribution of agro-pastoralism to rural well-being and economic development, its impact on the delicate balance of grassland ecosystems remains unclear in these crucial Tanzanian landscapes. The chapter aims to illuminate agro-pastoralism's environmental, social, and economic dimensions in these regions. Guided by research questions exploring current conditions, potential solutions, and the path toward sustainable grassland resource utilization, the study employed a systematic literature review and data analysis using R software. Key findings highlight challenges from the progressive expansion of agro-pastoral activities, leading to trade-offs between ecosystem services and productivity. The study identifies agro-pastoral clusters across the area, revealing variations in economic activities and their impact on grassland utilization. Impacts on natural resources, such as soil pH changes, reduced herbaceous biomass, and shifts in plant composition, are discussed. The legal framework related to natural resource conservation in grasslands emphasizes the need for a balanced, ecologically sustainable approach. Efforts to alleviate agro-pastoral impacts, including introducing climate-smart agriculture, are explored. The chapter concludes by emphasizing the importance of integrated, participatory methods for sustainable management in the Serengeti and Ugalla ecosystems. Recommendations include promoting sustainable land use practices, implementing rotational grazing, and enhancing community involvement in decision-making.

Keywords: agro-pastoral activities, grassland ecosystems, grassland utilization, grazing land, herbaceous plants

1. Introduction

Agro-pastoralism has been extensively studied at regional and global scales in relation to environmental processes [1]. However, the effect of agro-pastoralism on grassland ecosystems and the livelihood of people in Tanzania has remained unknown. It is widely acknowledged that agro-pastoralism contributes to the well-being of people in rural areas and the country's economy. On the other hand, little has been done to quantify the contribution of agro-pastoralism to the livelihood of people in Serengeti and Ugalla grassland ecosystems. The present study investigates the impact of agro-pastoralism on herbaceous plants and soil properties in Serengeti and Ugalla grasslands, as well as its contribution to livelihoods.

Grassland covers more than 40% of the Earth's surface [2] and about 6.8% of the land surface of Tanzania [3]. Grassland plays an essential role in ecosystem productivity and biogeochemical processes [4]. Grassland ecosystems have local importance for maintaining biodiversity and food production, influencing ecological processes, water, and climate regulation [5]. Approximately 49.3% of grassland in the world encounters degradation problems [6], and the grasslands of Tanzania are not exceptional in degradation problems. There is a progressive growth in human population and conversion of grasslands to agro-pastoral land [7]. This affects grassland as there can be conflicting goals between agro-pastoralism and the maintenance of grassland ecosystem services. Maximizing the provisioning of services from agro-pastoralism tends to result in a trade-off with plant composition and other ecosystem services. As a result of the non-linear relationships between ecosystem services, such as grassland composition and agro-pastoral productivity, managing them poses challenges [8]. When one service is optimized, other services are reduced or lost (a situation known as 'trade-offs') [8]. As communities continue to transform grassland ecosystems to obtain greater provision of specific services, the demand for more agro-pastoral land may lead to diminishing some grassland services.

1.1 Research questions

i. What is happening in the grasslands of Tanzania?

ii. Is there any solution to the negative impacts caused by anthropogenic activities?

iii. What is the way forward to sustainable utilization of available grassland resources in Tanzania?

1.2 The objective of the chapter

To depict the current situation of grasslands resulting from anthropogenic activities and its future scenario.

2. Methodology

2.1 Study area

The study was conducted in Western Serengeti and the Ugalla ecosystem in Western Tanzania (**Figure 1**). Western Serengeti is home to agri-pastoralists, and the population

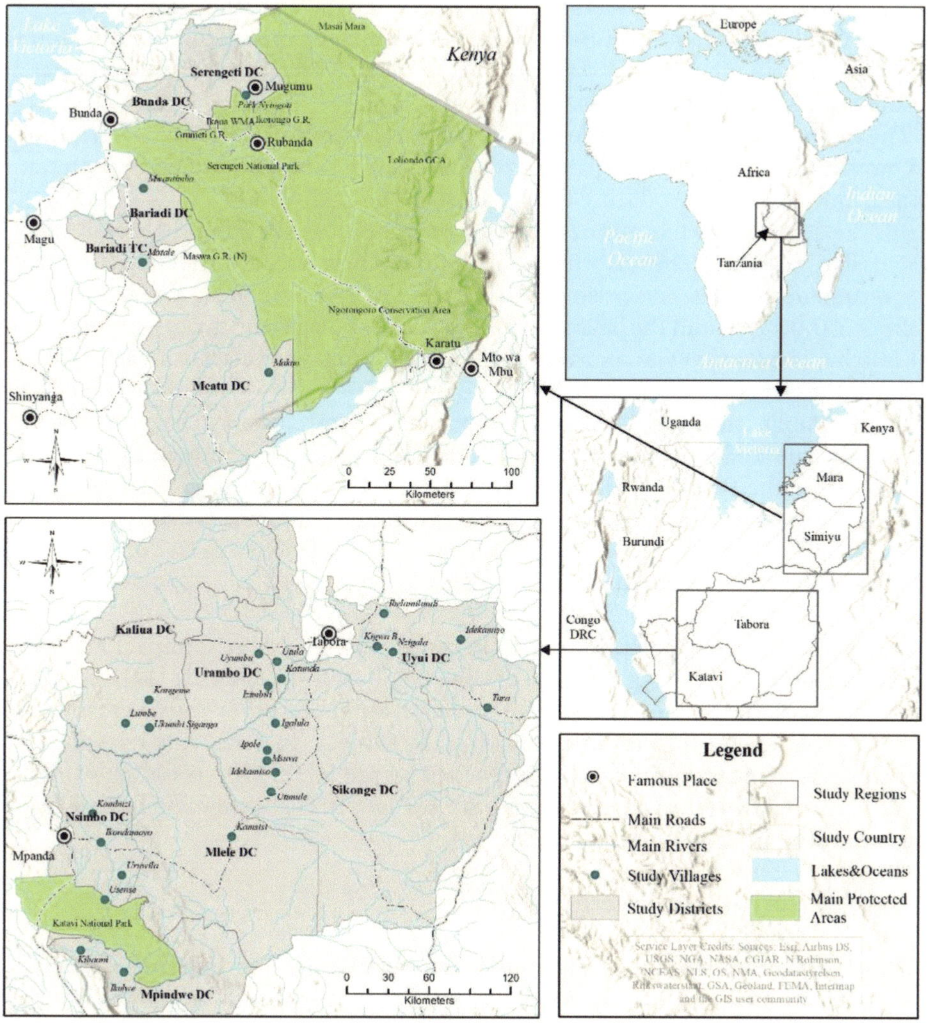

Figure 1.
Location of study sites in Serengeti and Ugalla ecosystems.

growth rate there is higher than in the north, east, and south of the National Park [9]. It is considered to have low suitability for arable agriculture. Although its subsistence economy is primarily based on agro-pastoralism [10], poor delivery of agricultural extension services, and insufficient input supplies limit this activity, which leads to the practice of extensive cropping and livestock keeping in villages that encroach on protected areas [11]. The study was conducted in four districts: Serengeti, Bunda, Meatu, and Bariadi.

Ugalla ecosystem comprises the Tanzania Ramsar site with many important socio-economic and cultural values. Some of the most important include harvesting wetland-related products, including fish, forest products, medicinal plants, honey, and wildlife. Other values of importance to the local communities include flood control, water supply, and dry-season grazing. The human population's subsistence economy in and around the proposed site depends mainly on farming, fishing, hunting, and honey gathering. Although honey gathering and fishing are not normally permitted in game reserves in Tanzania, it has been the practice to permit such activities in the game reserves in the Ramsar site as these activities predate the establishment of the reserves. Large numbers of fishing and beekeeping camps operate throughout the Ramsar site during the dry season (July to December). Permanent fishing villages are present around some of the lakes, such as Lake Sagara.

The estimated number of cattle is 15–20,000 in the central portion of the wetland and more cattle in the southern parts of the Ramsar site. Groups of nomadic pastoralists also move into the area during the dry season. The traditions of the local people on this site do not allow hunting or capturing of some birds like ground hornbill and animals like bushbuck. The majority of the Ramsar Site is under the direct jurisdiction of government agencies, comprising game reserves of 2.45 million ha, forest reserves of about 650,000 ha, and the balance being district or village land amounting to about 150,000 ha. The Ramsar site is surrounded by forest reserves in certain areas, notably in the northeast and south. In the central and northwest portions, the site is bounded by open or public land or agricultural areas primarily controlled by villages or district authorities. The study was conducted in five districts, including Urambo, Uyui, Kaliua, and Sikonge in the Tabora region and two districts in the Katavi region that include Mpanda (Nsimbo district council) and Mlele (Mlele and Mpindwe district councils).

2.2 Data collection and analysis

A systematic review of the scientific literature on agro-pastoral impact in grasslands was conducted using guidelines outlined by Pullin and Stewart [12] and Inskip and Inskip & Zimmermann [13]. Google Scholar, as well as other search engines, was used to determine the body of knowledge on the subject. To determine relevance and applicability, the search protocol was preceded by predefined filters for keywords [12, 13]. The relationship between agro-pastoral activities and edaphic factors was analyzed using R software version 3.5.0 [14].

2.3 Agro-pastoralism in Tanzania

Agro-pastoralism refers to a livelihood strategy that involves growing of crops and keeping of livestock by the local communities. This kind of livelihood significantly relies on rainfall patterns and the availability of natural pastures. Depending on cattle management types, there are three types of agro-pastoralism practiced in Tanzania. Brandström et al. [15] identified unilocal agro-pastoralism where livestock herds graze in the neighborhood of the homesteads and are taken back to craals every night. Another form of agro-pastrolism is bilocal, where herds graze near homesteads during certain months, but move far away for pasture and water during others. In addition, Brandström et al. [15] identified multilocal agro-pastoralism, in which a small number of cattle is permanently kept at the homesteads while the main herd is far from it throughout the year. In Tanzania, livestock density is highest in agro-pastoral areas around Lake Victoria [15]. This agro-pastoral area uses most of its labour in cultivation, where surplus is used to rear livestock by grazing crop residues. The same strategy is also applied in the agro-pastoral regions of central Tanzania [16]. Data from the 2014–2015 and 2016–2017 annual agriculture censuses [17, 18] show that agro-pastoralism is practiced throughout Tanzania's mainland (**Figure 2**).

Based on the data, the Mwanza region had the highest number of agro-pastoral operators whereas the Dar-es-Salaam region had considerably fewer operators. Further, cluster analysis revealed three main clusters of the regions practicing agro-pastoralism in Tanzania's mainland (**Figure 3**).

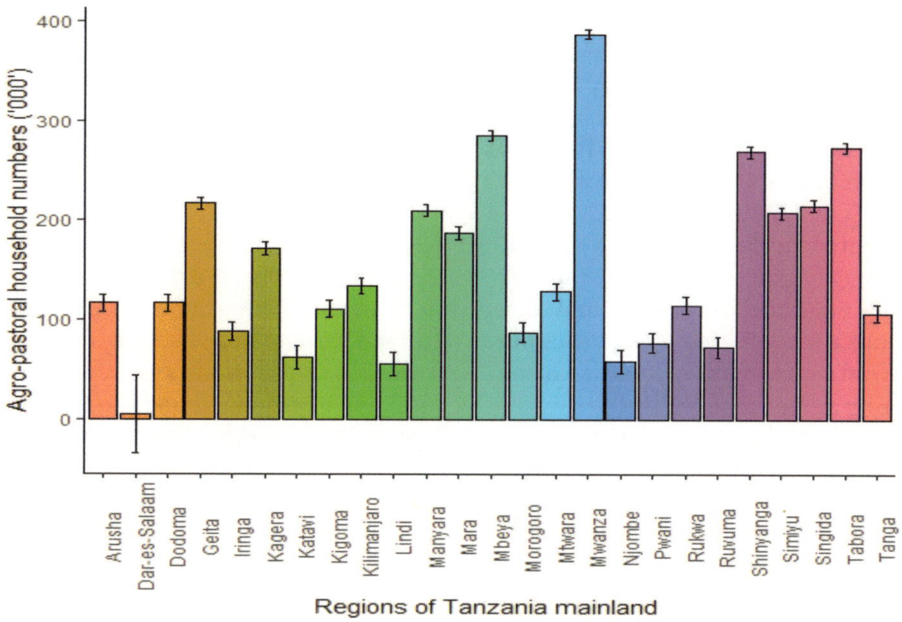

Figure 2.
Average number of agro-pastoral operators in different regions of Tanzania mainland. Source: Authors' synthesis from NBS [17, 18] data.

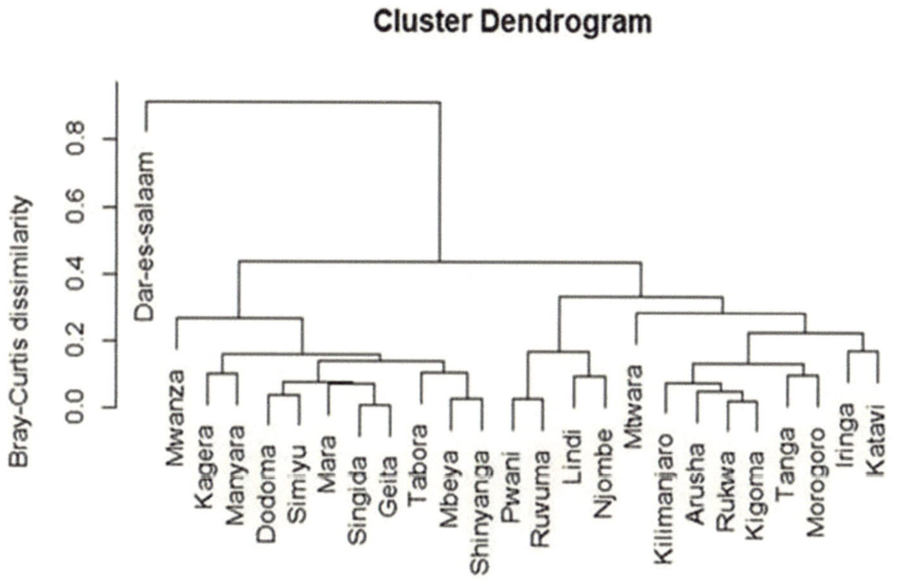

Figure 3.
Regional clusters of agro pastoral operators in Tanzania's mainland. Source: Authors' analysis from NBS [17, 18] data.

The Dar-es-Salaam region was a distinctive cluster. The other cluster included the regions of Dodoma, Mara, Singida, Mwanza, Kagera, Manyara, Simiyu, Geita, Tabora, Mbeya, and Shinyanga. The final cluster included the regions of Tanga, Morogoro, Lindi, Njombe, Pwani, Kilimanjaro, Arusha, Rukwa, Ruvuma,

Mtwara, Kigoma, Iringa, and Katavi. Additional research into the various clusters revealed that clusters 1 and 3 were dominated by crop farming (**Figure 4**). However, in cluster 1 operators were very few compared to other clusters. The reason for the low involvement of operators in cluster 1 can be associated with its city dwellers who are mainly engaged in salary jobs. Pastoralism activity was lacking in cluster 1, a circumstance implying the absence of land for roaming animals in the city.

In cluster 2, crop farming and agro-pastoralism businesses were fairly evenly distributed. In cluster 2, agro-pastoralism appeared to be a clear and unchangeable decision regarding social and economic life. As a result, the agro-pastoralists in cluster 2 likely descended from a pastoral society that turned to farming as a means of subsistence after suffering cattle losses due to natural disasters. Crop farming alone was the primary source of income in cluster 3. Cluster 3 agro-pastoralists likely arose from farming groups that decided to start raising cattle. Climate change and the requirement to use animals to boost agricultural output to obtain surplus crops in favorable years may drive the shift to livestock husbandry.

Compared to the Agrarian households in clusters 2 and 3, which engaged in crop farming and agro-pastoralism, fewer practiced pastoralism. This suggests that practicing pastoralism in Tanzania is challenging due to the effects of population expansion, climate change, and variability. Climate stress alters the amount, patterns, and distribution of rainfall and lengthens dry spells and droughts that reduce grazing land and water resources, resulting in decreased cattle output [19]. The number of households engaged in pastoralism in clusters 2 and 3 may decline if the country's climate stress persists.

Agro-pastoralism, in general, entails using tools, methods, and knowledge to alter the natural environment to create a variety of goods. Specific agro-pastoral clusters are shaped by the integration of economic activities used in agro-pastoral systems, and variances between these economic activities cause inconsistencies between clusters. Agro-pastoralism has criticism, and there are many diverse varieties that, depending on the economic circumstances of the local community, may place more or less emphasis on agriculture. The social, political, ecological, and geographic context in which these groups are found determines how often there are continual changes and transitions between more farming and more herding across time [20].

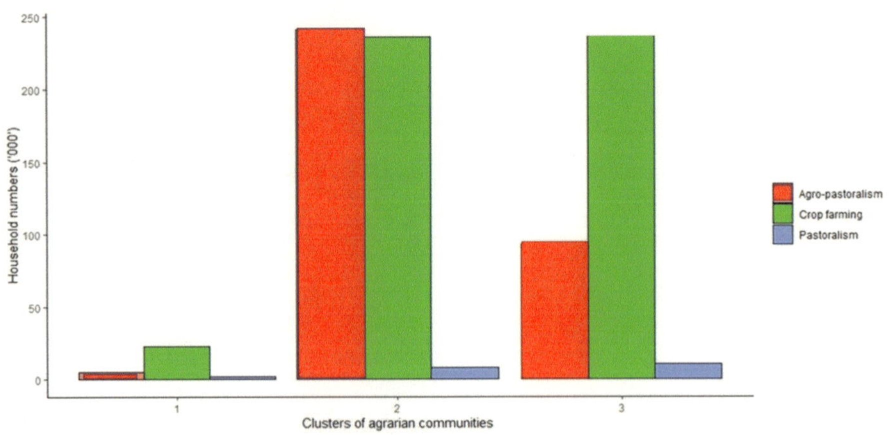

Figure 4.
Distribution of households involved in different agricultural activities among clusters in Tanzania's mainland. Source: Authors' synthesis from NBS [17] data.

2.4 Natural resources available in grasslands

Grasslands in Tanzania are dominated by native plants that include grasses, sedges, herbs, and scattered trees. The dominant plants in western Serengeti include *Themeda triandra* in wildlife grazing areas, *Cynodon dactylon* in livestock grazing areas, and *Chloris pycnothrix* in areas with mixed grazing of livestock and wildlife [21]. Grasslands in the Ugalla ecosystem are dominated by plant species that include *Digitaria macroblephara*, *Setaria pumila*, and *Echnocloa pyramidalis* in protected areas. In contrast, *Pennisetum polystachyon* commonly dominates cultivated land in villages (**Figure 5**).

Common trees in western Serengeti include *Kigelia africana* found along seasonal river banks; *Phoenix reclinate* found along rivers and swamps; *Commiphora Africana* scattered in various places; *Vachellia xanthophloea* found along rivers, swamps, and floodplains; *Vachellia tortilis* found in regular plains and *Vachellia drepanolobium* found in seasonally water-logged soils. Thornless deciduous trees adapted to uni-modal rainfall patterns are common in the Ugalla ecosystem. The common tree species include *Brachystegia augustistipulata*, *Brachystegia boehmii*, *Brachystegia glaberrima*, *Brachystegia glaucescens*, *Brachystegia longifolia*, *Brachystegia spiciformis*, *Brachystegia taxifolia*, *Julbernardia globiflora*, *Julbernardia paniculate*, *Isoberlinia angolensis*, and *Isoberlinia tomentosa*. The dominant genera in this ecosystem include Brachystegia, Julbernadia, Isorberlinia, Markhamia, Grewia, Terminalia, Syzygium, Vachellia, and Combretum [22]. Vegetations available in the grassland ecosystems of Ugalla and Serengeti provide non-timber forest products such as wild fruits, mushrooms, ropes, honey, bamboo, fodder, and brooms. Gathering of non-timber forest products helps local communities to boost their livelihood and households' upkeep. These products are used as food, medicine, and some of them are used for decorations.

2.5 Effect of agro-pastoralism on natural resources

Agro-pastoralism is a predominant livelihood in Serengeti and Ugalla ecosystems, where land and natural pasture are primary natural resources. Livestock moves in the vicinity of the villages and grazes on communal grazing lands, fallow lands, and crop fields after harvest. Studies showed that there is growth in human population and conversion of land to agriculture in agro-pastoral areas of Tanzania [7]. The increase in human population triggered the need for food sufficiency that stimulated cultivation and livestock keeping (**Figure 6**) based on land availability. Competition for

Figure 5.
Plants field assessment in Ugalla and Serengeti ecosystems.

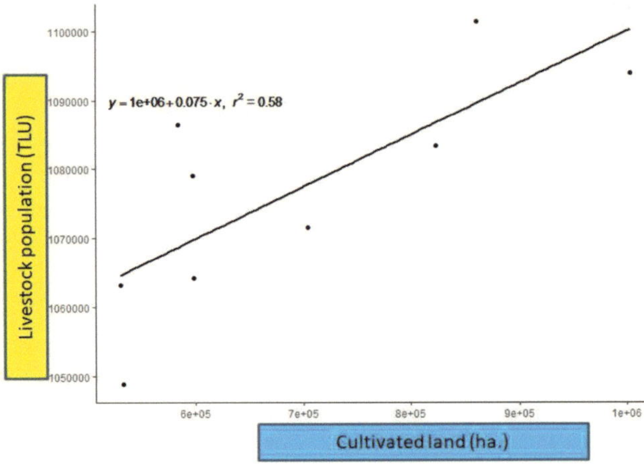

Figure 6.
Relationship between livestock population and land cultivated for crop production. TLU = Tropical Livestock Unit. Source: Authors' synthesis from NBS [17, 18] data.

natural resources, especially land, has become an issue of major concern and cause of conflict worldwide [23].

The change in land use in the upper catchments within ecosystems resulted in higher peak water flow during the rains and lower peak water flow during river drought [24]. Water abstraction levels for consumptive uses are normally highest during the dry season, imparting additional strain on the rivers within ecosystems during critical low water flow. This situation compelled the development of the Tanzania National Water Policy, which requires the protection of reserve flows to enable water availability for basic human needs and sustain ecosystems [25].

Soil is an important component of any ecosystem. It supports life systems by delivering ecosystem goods and services. For instance, the soil is one of the major carbon sinks, and plays roles in water regulation, soil fertility, and food production, which have effects on human well-being [26]. In any biological ecosystem, soil pH greatly influences biogeochemical processes [27]. Soil pH influences varieties of soil biological, chemical, and physical properties and processes that affect plant growth and biomass yield [28]. Despite the roles played by soil pH in ecosystems, keeping large numbers of livestock in communal grazing land tends to lower it [29]. For example, Tamartash et al. [30], demonstrated soil acidity to increase with increasing grazing intensity. Nevertheless, low soil pH affects soil microbial activities of decomposing organic carbon, shown by an increase in C:N ratio as the stocking rate increases (**Figure 7B**). Soil pH controlled microbial activity that consequently influenced microbial decomposition of carbon and nitrogen [27]. Therefore, agro-pastoral activities are envisaged to lead to acidification of soil.

Low soil pH (<6) in western Serengeti caused by agro-pastoral activities resulted in low herbaceous standing biomass (< 2000 kgDM/ha) and low herbaceous plant species richness (<5 species/quadrat), as shown in **Figure 6C** and **D**, respectively. Results obtained in western Serengeti concur showed that a soil pH of 6.3 was optimum for standing biomass and species richness of herbaceous vascular plants in miombo woodlands [31].

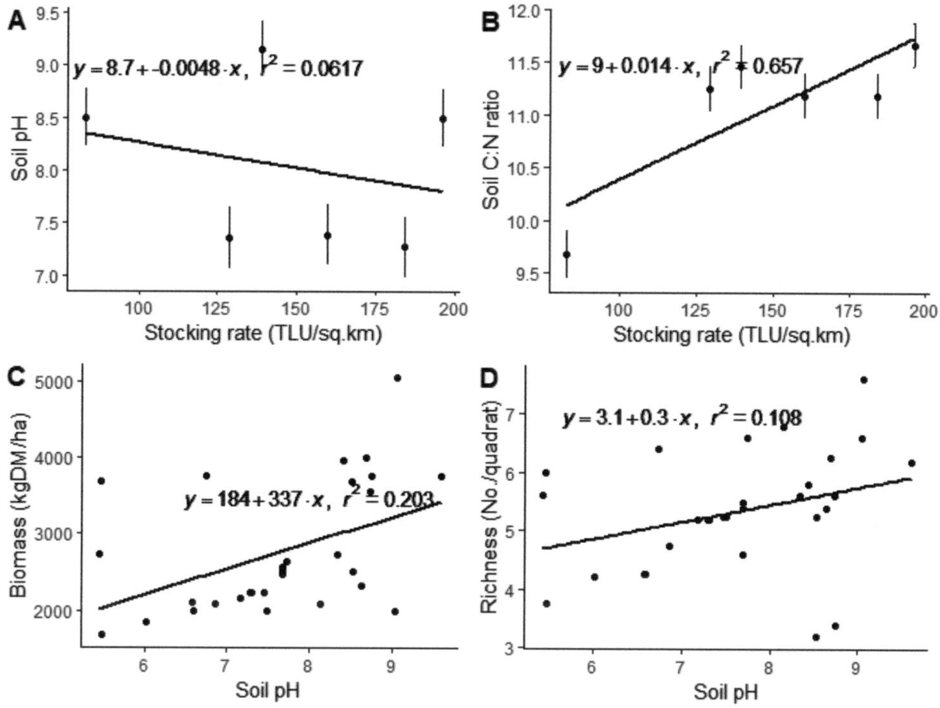

Figure 7.
Effect of animal density on soil properties and consequent effect on herbaceous plants attributes.

2.6 Legal framework on natural resources conservation in relation to grasslands

Wildlife Conservation Act No. 5 of 2009: The Act provides for the conservation of wildlife and ensures protection, management, and sustainable utilization of wildlife resources, habitats, ecosystems, and the non-living environment supporting such resources, habitats, or ecosystems with actual or potential use or value. The Act is important in grassland conservation as wildlife habitats in the sense that grasses form the main basal diet for wild herbivores in Tanzania.

Plant Protection Act No. 13 of 1997: The Act provides for the prevention of the introduction and spread of harmful organisms, to ensure sustainable plant and environmental protection, to control the importation and use of plant protection substances, and to regulate the export and import of plants and plant products. The Act is important in grassland conservation as it inhibits the introduction and spread of organisms that may alter grassland species' composition and ecosystem functions.

Land Act No. 4 of 1999 and Village Land Act No. 5 of 1999: The Land Act and Village Land Act have provisions that are important for environmental management. The fundamental principle of the Land Act is to ensure that land is used productively and that any such use complies with the principles of sustainable development. Among others, the Act prohibits any development activities in environmentally sensitive areas such as wetlands and swamps and 60 m from the shoreline and riverbanks. The Village Land Act also empowers the village government to have legal control of village land and its uses. This also includes prohibiting or minimizing land problems like bushfires as well as land use-related conflicts between farmers and livestock keepers/pastoralists. This Act limits human activities that may be detrimental to grassland ecosystems; for example, most of the grasslands in the Ugalla ecosystem are found in seasonally flooded lands. These areas are regarded by the Land Act as sensitive areas that require users to comply with sustainable development principles.

Water Resource Management Act No. 11 of 2009 and Water Supply and Sanitation Act No. 12 of 2009: The Water Resources Management Act (WRMA) provides the legal framework for the management of water resources within the integrated water resources management (IWRM) framework. The Act provides for pollution control and issues discharge permits for effluents to water bodies, including the underground strata. The Act also provides flood mitigation and control measures to inhibit or lessen the risk of flooding, flood damage, and water pollution. The Water Supply and Sanitation Act provides a legal framework to ensure water quality by protecting water works and storage facilities against pollution. The Act further gives a mandate to the Local Government Authorities to enact by-laws concerning water supply and sanitation for efficient and sustainable provision of these services in their areas. Grasslands are considered to regulate water above groundwater quantity and flow and limit soil erosion that contributes to water quality by minimizing sedimentation in rivers and lakes. This conquers the Act that emphasizes flood mitigation and pollution control.

2.7 The Grazing-Land and Animal Feed Resources Act (Act No. 13 of 2010)

This law sets the necessities to manage and control grazing lands, animal feed resources, and trade. The law establishes the National Grazing-land and Animal Feed Resources Advisory Council, which is responsible for promoting participatory, equitable use and management of grazing-land resources. The council works to foster collaboration with public and private institutions or authorities in issues related to the use and management of grazing land. Since most of the grazing lands are grasslands, this Act is concerned with the sustainable use of grasslands in a broad sense. However, it is bound to grazing lands that leave out grasslands with no grazing value but are important for the provisioning of ecosystem services.

2.8 Impacts of agro-pastoralism on grassland natural resources

Sustainable standing plant biomass production to support livestock and wildlife in grasslands requires understanding critical variables in soil, plant, and grazing effects imparted by grazers. A study conducted in western Serengeti showed that agro-pastoral activities resulted in low residue standing biomass availability in communal grazing and cultivated lands [3]. High livestock grazing pressure was the leading cause of low-standing biomass in communal grazing lands. It was shown that an increase in the stocking rate of grazing animals caused a decline in plant biomass. Heavy grazing due to a high stocking rate removes the growing points of grazed plants [32], thereby reducing plants' growth potential and, thus, a decline in plant biomass. Overgrazing and continuous cultivation in grasslands contribute to the deterioration of soil properties [29]. Agro-pastoral systems in the Western Serengeti and Ugalla ecosystems are mainly rain-fed.

The productivity of grasslands depends on the availability of water, either for plant growth or for livestock and wildlife production. Water for crop production and pasture growth comes from rainfall, while livestock and wildlife drink water from rain-fed water sources. It has been observed that streams and natural water sources have been shrinking due to rainfall variability and dry spells in ecosystems. Agricultural expansion, human population growth, and bushfires are considered to be salient ecological drivers of changes in the abundance of plants [33]. Agricultural expansion and human population growth are common phenomena in Western Serengeti and Ugalla ecosystems [3, 34, 35]. This situation leads to the shrinkage of grassland due to land clearance for cultivation. Therefore, the shrinkage of grassland and its product results in a consequential effect that devastated the sustainability of agro-pastoralism. The sudden effect of overgrazing and cultivation on the remaining part of grassland results in soil erosion, which affects the re-growth of plant and grass species. The pressurized grassland then starts to offer ecosystem services at a declining rate which results in a reduction in productivity.

2.9 Efforts to alleviate impacts of agro-pastoralism in Serengeti and Ugalla ecosystems

The performance of agro-pastoral systems is determined by the availability of land, water, and energy, which hit hard by climatic perturbation. In an endeavor to ameliorate the impact of agro-pastoralism and climate change in grasslands, the government, in collaboration with NGOs tried to introduce climate-smart agriculture. Climate-smart agriculture intends to increase productivity and income, the ability to adapt and build community resilience to climate change, and enhance food and nutrition security while achieving mitigation co-benefit in line with national economic development priorities. However, the adoption rate ranges from 4–30% [36, 37]. It was observed that climate-smart agriculture is site-specific that needs to consider local factors and co-design solutions with the communities that make use of grasslands.

3. Conclusion

Community livelihoods within the Serengeti and Ugalla grassland ecosystems are influenced by agriculture and livestock keeping, known as agro-pastoralist farming. Despite this, the study showed that agro-pastoralism adversely affects grasslands' natural resources. Plant composition, soil properties, and grassland ecosystem services are all affected by the continuous expansion of cultivated land and the keeping of large herds of grazing animals within limited grazing areas. Due to overgrazing and cultivation, soil fertility and plant biomass have been reduced, resulting in insufficient feed for grazing animals. As grasslands and natural resources are under pressure, sustainable grassland use is at risk.

3.1 Way forward

Integrated and participatory approaches are needed to ensure the sustainability of the Serengeti and Ugalla ecosystems. In order to preserve grasslands, agro-pastoralism policies, and management strategies should balance economic benefits and grassland ecology conservation. Emphasis should be placed on promoting sustainable land use practices that consider the land's carrying capacity and grasslands' ecological needs. Using rotational grazing, climate-smart agriculture, and soil conservation strategies can mitigate the adverse effects of agro-pastoral practices. Furthermore, involving local communities in decision-making will enhance a sense of ownership and stewardship, resulting in better community involvement in the management of natural resources. These interventions' success depends on the cooperation of government agencies, NGOs, and local communities. These holistic approaches can help to maintain the delicate balance between human livelihoods and preserving grasslands' biodiversity and ecosystem services.

Declaration

Some parts of this chapter are based on an unpublished part of the main author's Ph.D. thesis on 'influence of agro-pastoralism on herbaceous plants diversity and livelihood of communities in western Serengeti' submitted to the Senate of Sokoine University of Agriculture in Tanzania.

Author details

Pius Yoram Kavana[1*], Bukombe John Kija[2], Emmanuel Pagiti Reuben[1], Ally Kiyenze Nkwabi[1], Baraka Naftal Mbwambo[1], Simula Peres Maijo[1], Selemani Rehani Moshi[1], Shabani Matwili[1], Victor Alexander Kakengi[2] and Stephen Justice Nindi[2]

1 Tanzania Wildlife Research Institute, Western Wildlife Research Centre, Kigoma, Tanzania

2 Tanzania Wildlife Research Institute, Arusha, Tanzania

*Address all correspondence to: pykavana@gmail.com

References

[1] Dong S, Wen L, Liu S, Zhang X, Lassoie JP, Yi S, et al. Vulnerability of worldwide pastoralism to global changes and interdisciplinary strategies for sustainable pastoralism. Ecology and Society. 2011;**16**(2):10. Available from: http://www.ecologyandsociety.org/vol16/iss2/art10/

[2] Horion S, Cornet Y, Erpicum M, Tychon B. Studying interactions between climate variability and vegetation dynamics using a phenology based approach. International Journal of Applied Earth Observation and Geoinformation. 2013;**20**:20-32

[3] Kavana PY, Bukombe JK, Kija H, Nindi S, Nkwabi A, Lipende I, et al. Interaction of grassland ecosystem with livelihood and wildlife sustainability: Tanzanian perspectives. In: Grasses and Grassland-New Perspectives. London, UK: IntechOpen; 2021

[4] Lemaire G, Franzluebbers A, de Faccio Carvalho PC, Dedieu B. Integrated crop-livestock systems: Strategies to achieve synergy between agricultural production and environmental quality. Agriculture, Ecosystems & Environment. 2014;**190**:4-8

[5] Bengtsson J, Bullock J, Egoh B, Everson C, Everson T, O'connor T, et al. Grasslands—More important for ecosystem services than you might think. Ecosphere. 2019;**10**(2):e02582

[6] Gang C, Zhou W, Chen Y, Wang Z, Sun Z, Li J, et al. Quantitative assessment of the contributions of climate change and human activities on global grassland degradation. Environmental Earth Sciences. 2014;**72**:4273-4282

[7] Estes AB, Kuemmerle T, Kushnir H, Radeloff VC, Shugart HH. Land-cover change and human population trends in the greater Serengeti ecosystem from 1984-2003. Biological Conservation. 2012;**147**(1):255-263

[8] Rodríguez JP, Beard TD Jr, Bennett EM, Cumming GS, Cork SJ, Agard J, et al. Trade-offs across space, time, and ecosystem services. Ecology and Society. 2006;**11**(1)

[9] Kideghesho JR. 'Serengeti shall not die': Transforming an ambition into a reality. Tropical Conservation Science. 2010;**3**(3):228-247

[10] Emerton L, Mfunda I. Making wildlife economically viable for communities living around the Western Serengeti, Tanzania. Citeseer. 1999

[11] Mfunda IM, Røskaft E. Wildlife or crop production: The dilemma of conservation and human livelihoods in Serengeti, Tanzania. International Journal of Biodiversity Science, Ecosystem Services & Management. 2011;**7**(1):39-49

[12] Pullin AS, Stewart GB. Guidelines for systematic review in conservation and environmental management. Conservation Biology. 2006;**20**(6):1647-1656

[13] Inskip C, Zimmermann A. Human-felid conflict: A review of patterns and priorities worldwide. Oryx. 2009;**43**(1):18-34

[14] Dennis M, Saleh J, Snay R, Pearson C. HTDP User Guide (Software Version 3.5. 0). 2022

[15] Brandström P, Hultin J, Lindström J. Aspects of agro-pastoralism in East Africa. Nordiska Afrikainstitutet. 1979

[16] Rigby P. Cattle and Kinship among the Gogo: A semi-Pastoral Society of central Tanzania. (No Title). 1969

[17] NBS. 2014/15 Annual Agricultural Sample Survey Report. 2016. Available from: www.nbs.go.tz/nbs/takwimu/Agriculture/Annual_Agricultural_Sample_ Survey_Report2014_15.pdf] [Accessed: February 4, 2021]

[18] NBS. 2016/17 Annual Agriculture Sample Survey: Crop and Livestock Report. 2018. Available from: www.nbs.go.tz/nbs/ takwimu/Agriculture/Annual_ Agricultural_Sample_Survey_Report2016_ 17.pdf.] [Accessed: February 4, 2021]

[19] Sangeda AZ, Maleko DD, Mtengeti EJ. Socio-economic and ecological dimensions of climate variability and change for agro-pastoral communities in central Tanzania. Livestock Research for Rural Development. 16 Nov 2023;**25**(209). Available from http://www.lrrd.org/lrrd25/12/sang25209.htmLivestock

[20] Lane K. Through the looking glass: re-assessing the role of agro-pastoralism in the north-central Andean highlands. World Archaeology. 2006;**38**(3):493-510

[21] Kavana PY, Sangeda AZ, Mtengeti EJ, Mahonge C, Bukombe J, Fyumagwa R, et al. Herbaceous plant species diversity in communal agro-pastoral and conservation areas in western Serengeti, Tanzania. Tropical Grasslands-Forrajes Tropicales. 2019;**7**(5):502-518

[22] Banda T, Mwangulango N, Meyer B, Schwartz MW, Mbago F, Sungula M, et al. The woodland vegetation of the Katavi-Rukwa ecosystem in western Tanzania. Forest Ecology and Management. 2008;**255**(8-9):3382-3395

[23] Young H, Behnke R, Sulieman H, Robinson S, Mohamed A. Risk, Resilience, and Pastoralist Mobility. Somerville, MA, USA; 2016 Available from: http://fic.tufts.edu/assets/TUFTS_1611_Risk_Resilience_mobility_V6_online.pdf

[24] Mati B, Mutie S, Gadain H, Home P, Mtalo F. Impacts of land-use/cover changes on the hydrology of the transboundary Mara River, Kenya/Tanzania. Lakes & Reservoirs: Research & Management. 2008;**13**:169-177. DOI: 10.1111/j.1440-1770.2008.00367.x

[25] URT. National Water Policy – July 2002. 2002

[26] FAO. Revised World Soil Charter. Rome, Italy: FAO; 2015

[27] Neina D. The role of soil pH in plant nutrition and soil remediation. Applied and Environmental Soil Science. 2019;**2019**:1-9

[28] Minasny B, Hong SY, Hartemink AE, Kim YH, Kang SS. Soil pH increase under paddy in South Korea between 2000 and 2012. Agriculture, Ecosystems & Environment. 2016;**221**:205-213

[29] Kavana PY, Mtengeti EJ, Sangeda AZ, Mahonge C, Bukombe J, Nindi S, et al. Repercussions of agro-pastoralism on soil condition in Western Serengeti, Tanzania. Earth. 2022;**11**(3):130-141

[30] Tamartash R, Jalilvand H, Tatian M. Effects of grazing on chemical soil properties and vegetation cover (case study: Kojour rangelands, Noushahr, Islamic Republic of Iran). Pakistan Journal of Biological Sciences: PJBS. 2007;**10**(24):4391-4398

[31] Schaffers AP. Soil, biomass, and management of semi-natural

vegetation–Part II. Factors controlling species diversity. Plant Ecology. 2002;**158**:247-268

[32] Lyons RK, Hanselka CW. Grazing and browsing: How plants are affected. Texas FARMER Collection. 2001

[33] Naah J-BS, Braun B. Local agro-pastoralists' perspectives on forage species diversity, habitat distributions, abundance trends and ecological drivers for sustainable livestock production in West Africa. Scientific Reports. 2019;**9**(1):1707

[34] Lobora AL, Nahonyo CL, Munishi LK, Caro T, Foley C, Beale CM. Efficacy of land use designation in protecting habitat in the miombo woodlands: Insights from Tanzania. bioRxiv. 2017; 117622

[35] Lobora AL, Nahonyo CL, Munishi LK, Caro T, Foley C, Beale CM. Modelling habitat conversion in miombo woodlands: Insights from Tanzania. Journal of Land Use Science. 2017;**12**(5):391-403

[36] Kassie M, Teklewold H, Jaleta M, Marenya P, Erenstein O. Understanding the adoption of a portfolio of sustainable intensification practices in eastern and southern Africa. Land Use Policy. 2015;**42**:400-411

[37] Kurgat BK, Lamanna C, Kimaro A, Namoi N, Manda L, Rosenstock TS. Adoption of climate-smart agriculture technologies in Tanzania. Frontiers in Sustainable Food Systems. 2020;**4**:55